High Performance Liquid Chromatography Fingerprinting Technology of the Commonly-Used Traditional Chinese Medicine Herbs

High Performance Liquid Chromatography Fingerprinting Technology of the Commonly-Used Traditional Chinese Medicine Herbs

Baochang Cai
Nanjing University of Chinese Medicine

Seng Poon Ong
Temasek Polytechnic, Singapore

Xunhong Liu
Nanjing University of Chinese Medicine

With Contributions From

Pengchi Zhang, Wee Wee Ng, Helen Li Han (*Singapore*),
Haidong Wang, Xu Zhang (*Nanjing*)

Chemical Industry Press

World Scientific

NEW JERSEY • LONDON • SINGAPORE • BEIJING • SHANGHAI • HONG KONG • TAIPEI • CHENNAI

Published by

World Scientific Publishing Co. Pte. Ltd.
5 Toh Tuck Link, Singapore 596224
USA office: 27 Warren Street, Suite 401-402, Hackensack, NJ 07601
UK office: 57 Shelton Street, Covent Garden, London WC2H 9HE

and

Chemical Industry Press
No. 13 Qingnianhu South Street
Dongcheng District
Beijing 100011
P.R. China

Library of Congress Cataloging-in-Publication Data
Chang yong zhong yao cai HPLC zhi wen tu pu ce ding ji shu. English
High performance liquid chromatography fingerprinting technology of the commonly-used traditional Chinese medicine herbs / Baochang Cai, Seng Poon Ong, Xunhong Liu ; with contributions from Pengchi Zhang ... [et al.].
p. ; cm.
Includes bibliographical references and index.
ISBN-13: 978-9814291095 (hardcover : alk. paper)
ISBN-10: 9814291099 (hardcover : alk. paper)
I. Cai, Baochang. II. Ong, Seng Poon. III. Liu, Xunhong. IV. Title.
[DNLM: 1. Drugs, Chinese Herbal--analysis. 2. Chromatography, High Pressure Liquid--methods. 3. Quality Control. QV 766]
615.321--dc23
2011040825

British Library Cataloguing-in-Publication Data
A catalogue record for this book is available from the British Library.

Typeset by Rachel Wong

Printed by FuIsland Offset Printing (S) Pte Ltd Singapore

Foreword

Fingerprinting of Chinese medicine can fundamentally reflect the overall characteristics of Chinese medicine, giving an integrated evaluation and revealing the intrinsic quality of Chinese medicine, fulfilling its role in revealing the overall product content. It also guarantees the safety, efficacy and controllability of Chinese medicine, enabling Chinese medicinal research to be more aligned with China's integrated, holistic views and approach. The emergence of the fingerprinting technique in Chinese medicine as a key technology is leading the industry's holistic progress. Its research applications and role in enhancing quality standards, guiding the standardized production of medicinal products, propelling new product research and speeding up the modernization and globalization of Chinese medicine are both vital and practical.

Fingerprinting research in Chinese medicine has become an important research arena and a hot topic in our country's fundamental research in Chinese medicine. Its research standards and applications continue to increase and broaden. Research trends have broadened to become multi-disciplinary, mutually influencing, progressing from single indicators to multiple indicators in integrated development, with the aim of establishing fingerprints that reflect the efficacy of Chinese medicine. High-performance liquid chromatography (HPLC) has the features of high resolution, rapid analysis, high sensitivity, good stability, good reproducibility, wide choice of mobile phase, wide variety of detectors, and re-usability of columns. HPLC has thus become one of the important tools in fingerprinting research in Chinese medicine. The HPLC fingerprinting technique has played a major role in quality assessment, standardizing cultivation, new product research and other areas in Chinese medicine, thus kick-starting the broadening applications in many experimental works.

In recent years, we have developed many experimental works and established the Jiangsu Province Chinese Medicine Fingerprinting Research Centre, which is responsible for "Chinese Medicine

Fingerprinting Research" (Jiangsu Province New Advanced Technology Industrial Project), "Chinese Medicine Holographic Fingerprinting Research Database" (Jiangsu Province High School Natural Science Research Planned Research Subject 03KJB360092) and other assisted projects. In the process of this research, some admirable results were obtained. The research results have been compiled, summarized, written, edited and published into this book to serve as reference material for quality research work in Chinese medicine. With the progress of future research, sequels will be published subsequently.

In the research process, we attempt to standardize the operations at each step, trying our best to make it unique, reproducible and operational in order to attain reliable and practical results. The test samples are authentic products obtained from local markets, with variant identification performed on them. In the preparation of test samples, different solvents for extraction, extraction methods and purification methods are repeatedly investigated. The optimized extraction methods for the preparation of these test samples are selected, in an attempt to reveal as much as possible of the sample composition in the chromatogram. The chromatographic conditions include optimization of the choice of columns, mobile phase, detectors and others. These selections are necessary to build the best conditions for chromatography such that the components of the test samples can be successfully separated. In other words there should be as many chromatographic peaks as possible, to reveal the intrinsic characteristics of the product, provide sufficient information on the fingerprint evaluation and confirm its quality. The analysis and evaluation of fingerprints is carried out based on fingerprint data. Important parameters and similarities are established for fingerprint comparison and analysis. This will help build practical and workable quantitative evaluation standards for HPLC fingerprints. For convenience and practicality, fingerprint analysis and evaluation in this book mostly made use of visual indicators, with opportunities for further improvements.

The experimental research and written and editorial work of this book received strong support from Jiangsu Education Department and Nanjing University's Chinese Medicine Department. The successful publication of this book is due in part to the warm support, responsible editing and meticulous design work of Chemical Industry Press. We hereby extend our sincere thanks to all.

Due to the short writing and editing time, mistakes in the book are inevitable. Where there are shortcomings and mistakes, we humbly appeal to all readers to give us feedback accordingly.

Cai Baochang
Liu Xunhong
July 2004, Nanjing

Preface

In this era, as humanity begins to return to nature, advocating natural methods of healing, Chinese medicine is attracting more and more attention. The "safe, effective and controllable" nature of Chinese medicine is internationally recognized and thus its modernization and globalization is a sure trend. With the rapid development of modern analysis techniques and their application in the systematic research of Chinese medicine, the emergence of the Chinese medicine fingerprinting technique as a key technology is leading the industry's holistic progress. The role of this technology in enhancing Chinese medicinal quality control standards, guiding the standardized production of Chinese medicinal products, propelling new product research and speeding up the progress of its modernization and globalization is both vital and practical.

The fingerprinting technique is a quality control model that uses modern-day chromatography involving color spectrum, light spectrum and wave spectrum as support technologies. It is an integrated and quantifiable identification technique. It is an effective quality control method to evaluate the authenticity, stability and consistency of Chinese medicine with the unique characteristic of having multiple components in different material groups. Fingerprinting of Chinese medicine can fundamentally reflect the overall characteristics of Chinese medicine, giving an integrated evaluation and revealing the intrinsic quality of Chinese medicine, fulfilling its role in revealing the overall product content. It also guarantees the efficacy of Chinese medicine, enabling Chinese medicinal research to be more aligned with China's integrated, holistic views and approach. In recent years, fingerprinting research in Chinese medicine has become an important research arena and a hot topic in our country's fundamental research in Chinese medicine. Its research standards and applications continue to increase and broaden. Research trends have broadened to become multi-disciplinary, mutually influencing, progressing from single indicators to multiple indicators in integrated development, with the aim of establishing fingerprints that reflect the efficacy of Chinese medicine.

High-performance liquid chromatography (HPLC) has the features of high resolution, rapid analysis, high sensitivity, good stability, good reproducibility, wide choice of mobile phase, wide variety of detectors, and re-usability of columns. HPLC has thus become one of the important tools in fingerprinting research in Chinese medicine. The HPLC fingerprinting technique has played a major role in quality assessment, standardizing cultivation, new product research and other areas in Chinese medicine, thus kick-starting the broadening applications in many experimental works. Professor Cai Baochang in recent years has used HPLC fingerprinting research in Chinese medicine to develop many experiments, obtaining encouraging results. Based on that research, he has written this book. The contents of the book are both complete and accurate and its methods are concrete. It facilitates the broadening applications of the HPLC fingerprinting technique in practical work. From quality evaluation, resource development to new product research and even to other quality-related research in Chinese medicine, this book will prove to be a valuable reference. Its publication has added a monumental work to the literature on fingerprinting in Chinese medicine.

Ren Jiquan
May 2005

Contents

General Introduction

1.1

Outline

Chinese medicine has been used for several thousand years to treat and prevent illnesses. Its superior clinical therapeutic efficacy has won universal praise. Today, as natural methods of healing are promoted, Chinese medicine is receiving worldwide attention. Ever since Chinese medicine was used in the treatment and prevention of illnesses, problems with the authenticity and quality assessment of Chinese medicine have also arisen. Over time, the methods, techniques and theories of assessing the quality of Chinese medicine have also gone through the phases of formation, development and continuous improvement. In the beginning, identification of Chinese medicine was mainly based on the personal experience of individuals. Later, this gradually developed into four major identification methods, namely, origin identification, traits identification, microscopic identification, and physical and chemical identification. Subsequently, many modern identification methods based on instrumental analysis appeared. As modern analytical techniques progressed by leaps and bounds, and research into the Chinese medicine system became increasingly established, the fingerprinting technique for quality control in Chinese medicine was developed. This has accelerated the modernization process of quality control in Chinese medicine.

Here, a fingerprint refers to the common peaks of chromatograms which can be used to characterize a Chinese medicine and are obtained through specific analytical methods. A fingerprint reflects the overall nature of a Chinese medicine, which elevates the quality control target from analysis of a single component to analysis of all the material in a

Chinese medicine. This ensures consistency in the clinical therapeutic efficacy of Chinese medicine, and makes research in this field more aligned with the holistic approach of Chinese medical science. Hence, this has been considered a milestone in Chinese medicine quality control. In recent years, fingerprinting technology has been increasingly applied in many areas such as quality assessment, resource development and the search for active ingredients. It has become one of the important means for variant identification and quality assessment of Chinese medicine. The establishment of a Chinese medicine fingerprint database provides a reference and guideline for routine testing and analytical work, and serves as a solid foundation for the quality standards of Chinese medicine.

Origin and Development of Fingerprinting

Concept of Fingerprinting

The fundamental difference between Chinese medicine and chemically synthesized drugs is that the effectiveness of a Chinese medicine depends on how various chemical components work together, while that of a chemically synthesized drug usually depends on only one chemical component. So the practice of evaluating traditional Chinese medicine based on the quantitative and qualitative analysis of a single chemical component is gradually being questioned in terms of its effectiveness and accuracy. This is because it is difficult to effectively evaluate the quality and authenticity of a Chinese medicine based on any single active ingredient. This is especially so if the analyzed ingredient is also a common ingredient in many Chinese medicines, thus further reducing the degree of accuracy and specificity of the identification process. In order to increase the accuracy and objectiveness of the identification process, modern analytical techniques are used to analyze the overall effective components of different herbs. One of the techniques, fingerprinting, is based on this concept and has been attracting more interest in recent years.

Fingerprinting is a quality control model that builds upon spectroscopic and chromatographic technology. It is different from the traditional quality control model in the sense that fingerprinting looks at the "complete information" or comprehensiveness of the chromatograph, and displays integrated quality information. Since the secondary metabolites, which are chemical components of medicinal herbs, are inherently unstable, the fingerprints of these chemicals possess a fuzziness that cannot be precisely measured, just like the fuzzy phenomenon in our daily lives. Comprehensiveness and fuzziness are the two basic traits of a fingerprint. The similarity of fingerprints is established through these basic traits. Fingerprint analysis focuses on accurate identification (of similar peaks), and not on precise calculation. The comparison of fingerprints emphasizes similarity and the fingerprints compared do not need to be exactly the same. When it is impossible to find out all the complex components of a Chinese medicine, fingerprints can be used to check the stability of the intrinsic quality of the medicine.

In summary, fingerprinting can be defined as a type of comprehensive and quantifiable method for authentication; it is one of the quality control models that are suitable for evaluating the authenticity and stability of Chinese medicine. Comprehensiveness and fuzziness are its basic traits. The fingerprints should be specific and reproducible and the techniques of the fingerprinting should be practical. This definition explains the essence of Chinese medicine fingerprinting. It shows that fingerprinting is an integrated and comprehensive method of identification. Not only does it differentiate authentic medicines from fakes, it can also evaluate the stability of the quality of a Chinese medicine. Based on its integrative and comprehensive nature, it can be used for accurate and quantifiable authentication and quality evaluation of Chinese medicine. Yet from the perspective of details, the results are fuzzy. Especially since Chinese medicinal herbs are influenced by factors such as the ecological environment and the time of collection, it is very difficult to provide a ratio of common peak to non-common peak that is applicable to all the test samples.

Evolution and Current Status of Fingerprinting

Evolution of fingerprinting

The Chinese medicine fingerprinting technique originates from the development of modern analytical techniques. In the 1970s, trials were carried out to use thin-layer chromatography (TLC) to analyze Chinese medicine. Due to the limitations of the conditions, a lack of maturity in the technique and the timing, it did not receive public acceptance and disappeared from the scene after a period of time. Thereafter, studies in this area were mainly restricted to academic research. In the traditional quality control model of Chinese medicine, there was no requirement for fingerprinting. Hence, it made no impact on the Chinese medicine manufacturing industry. Ever since the State Food and Drug Administration of China enforced the use of fingerprinting as a quality control method for Chinese medicine injection, it has drawn keen nation-wide attention from departments involved in Chinese medicine production, teaching, research and quality control. This generated a strong driving force for research in Chinese medicine fingerprinting.

Fingerprint analysis is an extension of the traditional authentication techniques. In the 1985 edition of *Pharmacopoeia of the People's Republic of China* (Part 1), besides authentication through physical and chemical methods, there was also a greater use of TLC for authentication. During that time, TLC authentication was all conducted in relation to chemical references, and the requirement was met when the sample showed the same spot as the reference material in the chromatograph. However, due to the limited number of chemical references available, authentication could not be carried out when such references were absent. In addition, to rely on one chemical reference only is not specific enough, especially for a chemical component that is common in many medicinal herbs. Hence, when the 1990 edition of *Pharmacopoeia of the People's Republic of China* was released, herb references were added for TLC authentication. This resolved to a certain degree the pressing problems of limited availability of chemical references and the non-specificity of using only one chemical reference. In addition, a herb reference in a chromatograph provides much more information than a single chemical reference. The change or addition of herb references, quantification of samples, comparison of whole chromatographs, and authentication of test samples based on whole chromatographs all possess the initial structure of fingerprinting. However, the process leaves room for improvement due to a lack of standardization of the test samples, test equipment and test procedures. And because of that, the concept of fingerprinting was also not established during that time.

Since the initial trials in the 1970s, and with the development of modern analytical techniques, subsequent academic research, the foundation of the widespread application of herb references for TLC authentication established in the 1990, 1995 and 2000 editions of *Pharmacopoeia of the People's Republic of China*, we have taken a step further in the standardization of medicinal materials, operating conditions and test methods. The concept of Chinese medicine fingerprinting has come into focus and should gradually develop as an extension of current Chinese medicine authentication. Also, through fingerprint research, the overall standard of Chinese medicine cultivation, collection, processing, production, analysis and quality control is elevated. The diversity and continuous improvement of the level of current instrumental analysis techniques have also provided a good working platform for fingerprinting research and applications in Chinese medicine.

Current research in fingerprinting

Fingerprinting has become an important domain and hot spot for current fundamental research in Chinese medicine in China. There are a good number of academic research reports. The content includes theoretical discussions of Chinese medicine fingerprinting, standardized testing for the analysis of fingerprints, research in the methodology of establishing fingerprints, informatization and knowledge discovery research, as well as the applications of fingerprinting in improving quality control standards, guiding standardized production, developing new resources and new formulations of Chinese medicine, etc. The research has obtained rather satisfactory results. In practical applications, the national Chinese medicine injection fingerprint research plan has been launched officially. More than a hundred specialists and scholars from 48 research institutions and institutions of higher learning will complete a study of 74 types of preparation involving 146 Chinese medicine injection production enterprises. The State Food and Drug Administration promulgated the Guidelines for the Test Procedures of Chromatographic Fingerprints of Chinese Medicine Injection and two types of computer-aided Chinese medicine fingerprint similarity computation software. The guidelines detail the requirements for test sample collection and preparation, selection of reference, fingerprinting test conditions, laboratory and instrumentation facilities, test method, and the construction, identification, verification and checking of chromatographic fingerprints for raw materials, medicinal materials, intermediates and final products. The two types of computation software are able to compare and compute all the Chinese medicine chromatographic peaks, to determine the overall degree of similarity between the fingerprints. They automatically process the information to accurately evaluate the stability of the quality of the Chinese medicine. This established a practical technical platform and scientific foundation for the comprehensive research of Chinese medicine fingerprinting. Currently, the study of Chinese medicine fingerprinting is still primarily at the initial stage, which focuses on determining the methods of obtaining fingerprints, determining the evaluation criteria, and evaluating and controlling the quality of Chinese medicinal material.

Outside of China, fingerprinting is listed in the US Food and Drug Administration's (FDA) Guidance for Industry Botanical Drug Products (drafted in year 2000), the World Health Organization's (WHO) Guidelines for the Assessment of Herbal Medicine (1996), the British Pharmacopoeia (1986), the Indian Pharmacopoeia (1998), and the US Pharmacopeia (1999–2001). The objectives of these guidelines are to resolve the problem of unknown active constituents in many herbal medicines and to address the problem of authenticating and controlling the quality of commercial products. They require the herbal medicine product manufacturers to provide the fingerprints of intermediates to ensure the authenticity of the species, and the fingerprints of final products to verify the batch-to-batch stability in quality. The US FDA allows the provision of chromatographic fingerprint authentication information for applications relating to herbal health products. In the WHO's Guidelines for the Assessment of Herbal Medicine, where the active constituents cannot be distinguished in the herbal medicine products and final products, provision of the fingerprints of characteristic constituents or mixture of constituents will ensure consistency in the product quality. The number of overseas research articles on fingerprinting is also growing. These articles recognize too that the efficacy of a particular herbal medicine cannot be attributed to a single constituent. In some herbs, the active constituent is not known, and is difficult to identify. Hence, fingerprint identification is brought up as a possible solution. Basically, the research and application of fingerprinting is currently restricted to monitoring and ensuring the stability of the inter-batch quality of herbal medicine with complex constituents and unknown active constituents.

Classification and Development of Fingerprinting

Classification of fingerprints

The fingerprint that we refer to presently is an extension of and development from the use of complete chromatograms of reference herbs to identify test samples. A fingerprint, in the narrow sense, refers to the fingerprint of the chemicals in a Chinese medicine, which represents the characteristics of the chemical metabolites in the medicinal herbs. A fingerprint in the general sense can be classified into different types according to the object of application or the method of analysis.

Classification according to object of application

Fingerprints can be used at every stage of the preparation research and manufacturing process. They can be classified into Chinese medicinal material (raw material) fingerprints, processed medicinal material (including sliced medicinal material, blended pellets) fingerprints, intermediate (intermediary product in the production process) fingerprints and preparation fingerprints, depending on the object of application.

Chinese medicinal material fingerprints can be further classified into chemical fingerprints and DNA fingerprints, according to the method of analysis. Chinese medicinal material DNA fingerprints are mainly for the determination of the DNA profile of each medicinal material. Due to the unique and hereditary nature of the genes of every species, DNA fingerprinting of Chinese medicinal materials can be used in species identification, botanical classification research and quality research. It is extremely useful in research on the setup of Good Agricultural Practices (GAP) bases, cultivation standards (standard operating procedures or SOP), selection of premium cultivation resources and the authenticity of Chinese medicinal herbs. Chinese medicinal material chemical fingerprints refer to fingerprints setup through the analysis of each chemical constituent (secondary metabolites) in the medicinal material. Although chemical constituents are secondary metabolites, which are affected by the biological environment and life span and so produce obvious differences in individual plants, the metabolism of plants is hereditary in nature and plants of the same species will possess similarities in their chemical composition. Hence, the chromatograms of the chemical constituents can be used to setup the fingerprints. The use of chemical fingerprints of Chinese medicinal materials in quality control is more direct and of greater significance.

Processed medicinal material fingerprints should be treated separately. The requirements of the fingerprints for sliced medicinal materials are basically the same as for the medicinal material itself. For blended pellets (including standard extracts), the fingerprint requirements should be greater. The standard extraction procedure should be carried out. Operating parameters should be controlled strictly. The method of mixed-batch blending should be used to process medicinal materials, so that it is completely feasible to establish a stable fingerprint for quality control.

In establishing stable chemical fingerprints for Chinese medicine preparations and intermediates, if one can change the feeding materials from raw materials to extracts (maybe using the mixed-batch blending method), enforce strict control over the process flow, and monitor the entire production process through fingerprinting, the problem of quality stability of the final product could be resolved.

Classification according to method of analysis

Chinese medicine fingerprints can be classified into chemical fingerprints and biological fingerprints according to the method of analysis. Chinese medicine chemical fingerprints refer to fingerprints obtained through chromatographic, spectroscopic and other methods of analysis to establish the characteristics of the chemical constituents. Generally, these include thin-layer chromatography (TLC) fingerprints, gas chromatography (GC) fingerprints, gas chromatography–mass spectrometry (GC-MS) fingerprints, high-performance liquid chromatography (HPLC) fingerprints, high-performance liquid chromatography–mass spectrometry (HPLC-MS) fingerprints, high-performance capillary electrophoresis (HPCE) fingerprints, high-speed counter-current chromatography (HSCCC) fingerprints, supercritical fluid chromatography (SFC) fingerprints, ultraviolet spectroscopy (UV) fingerprints, infrared spectroscopy (IR) fingerprints, fluorescence photometry (FP) fingerprints, nuclear magnetic resonance (NMR) fingerprints, mass spectrometry (MS) fingerprints, X-ray diffraction (XRD) fingerprints, and hyphenated chromatography fingerprints (multi-dimensional multi-data characteristic chromatograms). Besides these, there are electrochemistry fingerprints, electrophoresis (EP) fingerprints, differential thermal analysis (DTA) fingerprints, circular dichroism (CD) fingerprints, trace element fingerprints, X-ray fluorescence spectroscopy (XRF) fingerprints and oscillating fingerprints.

Chinese medicine biological fingerprints include DNA fingerprints, genomic fingerprints and proteomic fingerprints. Chinese medicine genomic and proteomic fingerprints refer to the use of Chinese medicine preparations to act on certain cells or animals, resulting in complex changes in the condition of the genome and proteins. These two types of fingerprints are also called bioactive fingerprints. Besides, there are scanning electron microscopy (SEM) fingerprints and computer image analysis (CIA) fingerprints.

Development of fingerprinting

Fingerprinting, as a breakthrough in the modernization of Chinese medicine, should not be restricted to the elevation of the quality standards of Chinese medicine. Instead, one should get to know fingerprinting from the vantage point of how the modern scientific system should be used with traditional Chinese medicine, and finally to work towards the globalization of Chinese medicine. Modern cross-disciplinary language could be used to describe the essence of Chinese medicine. After decades of research and development, copious research has been conducted on Chinese medicine (Chinese medicinal material, sliced medicinal material and Chinese proprietary medicine), accumulating a large amount of research data on the chemical compositions of single medicinal herbs, research approaches and partial data on the study of chemical compositions of Chinese medicine formulae. The rapid development of modern analytical science has seen the emergence of numerous advanced methodologies, stable and reliable equipment and technology for setting up fingerprints. The rapid development of hyphenated technology has, in particular, provided the possibility of online analysis of unknown components. The development of information science allows the huge amount of data obtained to be informatized. Complex systems science has provided us with an overall approach to and a corresponding method for resolving such a complex system as Chinese medicine. All these helped to establish a solid foundation for the large-scale development of Chinese medicine fingerprint research, for the establishment of standardized and functional fingerprints, and for the integrated evaluation of Chinese medicine.

Developmental phases of fingerprinting

The research and construction of Chinese medicine fingerprints can be classified into initial and advanced phases.

The initial phase should determine the fingerprint construction method and assess the degree of similarity. Chinese medicine injection is fast-acting, has good therapeutic effects, and its chemical constituents are comparatively fewer than other forms of Chinese medicine. Yet the requirements for product quality are more stringent compared to those of other dosage forms. It is highly feasible to choose Chinese medicine injection to proceed with the work of setting up fingerprints. In the case of Chinese medicine injection, for the initial phase, the fingerprints are analyzed through the injection forms (final products), intermediate products and raw medicinal materials. The corresponding indicative control parameters are setup, to achieve the aim of obtaining operable, controllable, stable and quantifiable fingerprints. Fingerprints can be used to represent the final product quality, to establish quality control for the production process and the medicinal materials, to trace the source of problems in the production process and to demonstrate the quality requirement for the GAP of the medicinal herb. After a large amount of fingerprint data has been obtained, the method and indicators for determining the degree of similarity among fingerprints can be discussed. One can use the "common peak" method to determine the degree of similarity for the purpose of control, or use pattern recognition to determine the degree of similarity among the overall fingerprint chromatograms, and determine the various control parameters. Research in the initial phase focuses on setting up the analysis method, investigating the method stability and suitability, basic research on the active chemical components, multi-indicator control of the fingerprint, etc, which elucidate the chemical basis of the Chinese medicine and the quality control parameters, and makes quality control of Chinese medicine quantifiable.

After a relatively complete Chinese medicine fingerprint has been setup, one enters the advanced phase of the research, which is the investigation into the relationship between fingerprint characteristics and drug efficacy, and the bioequivalence of fingerprints. This is the basic meaning of multi-dimensional and multi-data characteristic fingerprints and fingerprint pharmacodynamics. Western medicinal compounds have a quantitative structure-activity relationship, which means a certain molecular structure has a corresponding biological activity. Chinese medicine takes effect through a group of chemical compounds that have drug efficacy. Different ways of herb blending (including changes in the type and quantity of herbs) will result in changes in the type, number and quantity of chemical compounds. Such changes in chemical groups can be expressed through fingerprints. The correlation between the changes in chemical composition (type, number and quantity) and drug efficacy (pharmacodynamic studies or clinical effects) can be established through a fingerprint. At this point, the fingerprint obtained is not only the integral expression of the chemical composition; it also contains biological information on related drug efficacy. Hence, one is able to achieve corresponding drug efficacy and clinical effects though control of the fingerprint. For different Chinese medicinal herbs, the differences in fingerprints are even greater; the corresponding drug efficacies are also not the same. By making changes to the drug blending and conducting research on the quantitative correlation between chemical composition and drug efficacy, one is able to use fingerprints to express different biological information. If new drug development is able to make use of fingerprints, right from the start, to manifest the changes in the chemical groups, together with pharmacological and pharmacodynamic testings as well as correlation analysis, the development of new drugs will be sped up and a higher standard can be obtained.

There is no clear demarcation between the initial and the advanced phases. In the initial phase, the main objective is to carry out relatively comprehensive quality control. Hence, the research should be in-depth, and during execution, it should be simplified so as to facilitate its widespread use. The data accumulated during the initial phase provides the foundation for research in the advanced phase. Fingerprints that are able to express integrated information about the correlation between chemistry and drug efficacy will help to speed up the modernization of Chinese medicine, and will provide reliable quality assurance for the globalization of Chinese medicine.

Informatization and knowledge discovery of fingerprints

In the development of Chinese medicine fingerprinting, the advancement of information science can help the informatization and knowledge discovery of fingerprints. Fingerprint informatization includes data acquisition and digitization. The main process includes selecting a suitable analysis method to obtain the fingerprints, setting up the operating conditions for the entire analysis procedure, and conducting tests to obtain the fingerprints of various samples, which is known as data acquisition. The obtained fingerprints are analyzed to determine their numerical characteristics. For example, in infrared fingerprints, the peak numbers and peak intensities can be the numerical characteristics used to define the peaks. For HPLC fingerprints, the peak parameters such as the retention time (or relative retention time) and peak height (and/or peak area) can be the numerical characteristics. Direct observation of the actual fingerprints allows easy comparisons between the fingerprints under simple conditions, whereas extraction of the numerical characteristics to obtain a digitized fingerprint allows for ease of mass storage, comparison and utilization. Hence, a database should be setup for large volumes of fingerprints. For Chinese medicine, a fingerprint database for different variants, places of origin, times of collection and processing methods should be setup for ease of comparison, so as to determine the degree of variation. As for Chinese medicine formulation (Chinese proprietary medicine), besides a fingerprint database for the Chinese medicine formulation, a fingerprint database of the entire process including raw materials, intermediates or effective parts, and final products should be setup as well. This is to facilitate quality control in the entire process and to trace the effect of changes in operating conditions and raw materials on product quality. For new drug research and development, the fingerprints of research samples at each phase can be collected into a database, which will enable monitoring and characterization of the chemistry and drug efficacy of each effective part. The capacity to add methods and functions to evaluate and compare fingerprints to those in the existing database will render it an intelligent database. A sample fingerprint could then be keyed in and compared with existing fingerprints, and hence interpreted to determine its quality.

Fingerprint knowledge discovery refers to the study of how to utilize the information after the fingerprints have been informatized. Fingerprint knowledge discovery includes interpretation, comparison and evaluation of information, study of the correlation between chemistry and drug efficacy, and information utilization, which is the extraction of regular patterns and knowledge from a large volume of fingerprint data. China has launched some research work in this area. Chinese medicinal materials have been extracted and separated to obtain the constituent chemicals. Under the guidance of traditional Chinese medicine (TCM) theory, stepwise regression analysis and canonical correlation analysis are conducted on pharmacological and chemical data obtained from related pharmacological experiments, to confirm that a particular chemical constituent in the Chinese medicinal material is the main material basis for its effectiveness. The particular active constituent is chosen as the subject for study. The sample medicinal materials are extracted and separated to prepare sample solutions, and the fingerprints (such as HPLC fingerprints) of the active constituent are analyzed. The chromatographic peaks are obtained and used as the variables for chemical characteristics. Reference peaks are also used in fingerprint analysis. After the fingerprints are informatized, chemical pattern recognition is carried out to determine the quality of the medicinal materials. In the case of Chinese medicine preparation, based on the known therapeutic effects of the prescription, the corresponding pharmacodynamic test method is determined. Based on the current chemical and pharmacological information on the medicinal materials (or effective parts and constituents) that make up Chinese proprietary medicine, and production technology, the possible class of chemical constituents is determined. Modern separation techniques (such as supercritical fluid extraction, macroporous resin adsorption, counter-current chromatography and all types of preparative chromatography) are used to separate the sample into each chemical type (chemical group) for pharmacodynamic testing, in order to finally determine the effective parts (the chemical constituents that are verified by pharmacodynamic testing) in the Chinese medicine. Its fingerprint (such as a HPLC fingerprint) is established and the chemical information (including the number, content and ratio of chemical constituents) and drug efficacy information obtained are analyzed for overall correlation, to obtain the cause and effect relationship (namely the regular pattern and knowledge).

An intelligent Chinese medicine fingerprint database is a database that collates a large number of fingerprints from medicinal materials, intermediates and final products. The correlation of the peaks can be determined using mathematical comparison methods. This is such that total quality management can be realized for the entire production process. As for basic research and new drug development, the characterization of chemical constituents for the entire process can be actualized. In addition to multi-dimensional (HPLC-PDAD-MS/MS) fingerprints, a large amount of information needs to be processed for the entire research. Hence, setting up an intelligent fingerprint database will help to resolve this problem.

Developmental prospect of fingerprinting

The development trend of Chinese medicine fingerprinting is built upon its fingerprint pharmacodynamics, actualizing the informatization and knowledge discovery, making Chinese medicine safe, effective and controllable. In actual applications, there are several aspects in Chinese medicine fingerprint quality control technology that can be further developed:

- a rational, easy-to-implement fingerprint similarity analysis method and the corresponding software, that is packaged with the analytical instrument, to be used by Chinese medicine manufacturing companies and drug analysis departments for quality control and evaluation;
- Chinese medicine analysis, identification, classification and evaluation software and network platform to service all the relevant departments, companies and organizations;
- online fingerprint analysis technology, analysis software and production automation based on fingerprint information, so as to actualize the control of Chinese medicine production process using fingerprints;
- a fingerprint database and network platform for Chinese medicinal materials (raw materials, especially for authentic medicinal materials),

processed medicinal materials (including sliced medicinal materials, blended pellets), intermediates (intermediate products from the production process) and preparations, to ensure total quality management of Chinese medicine;

- analysis and evaluation software for the correlation of fingerprint characteristics and drug efficacy, in order to establish fingerprints with integrated information on the correlation between chemistry and drug efficacy, speed up the new drug development process and raise the standard of research.

In summary, Chinese medicine fingerprinting possesses promising developmental prospects and potential for important applications. It will definitely play an even more significant role in the modernization of Chinese medicine.

A schematic diagram of the origin and development of Chinese medicine fingerprinting is as shown in Fig. 1.

Significance and Applications of Fingerprinting

To date, the quality control technology for Chinese medicine fingerprinting has become vital in driving the overall improvement of this industry. Its application and research have been significant in strengthening the Good Laboratory Practice (GLP) of Chinese medicine, ensuring its efficacy, realizing the Good Manufacturing Practice (GMP) of Chinese medicine, raising the overall standard of the Chinese medicine industry, implementing the GAP of Chinese medicinal materials, leading the agriculture of Chinese medicine towards modernization, and driving Chinese medicine towards globalization.

Adoption of Fingerprinting Technology to Ensure the Efficacy and Modernization of Chinese Medicine

Chemical substances in Chinese medicine are very complex. Ginseng is the species with the most comprehensive and the longest history of research on its chemical composition. Its documentation can be traced back to 1854 in a report on ginseng methaqualone by Sarriquex S. in *Annalen der Chemie und Pharmacie*. In the past 30 years, new gin-

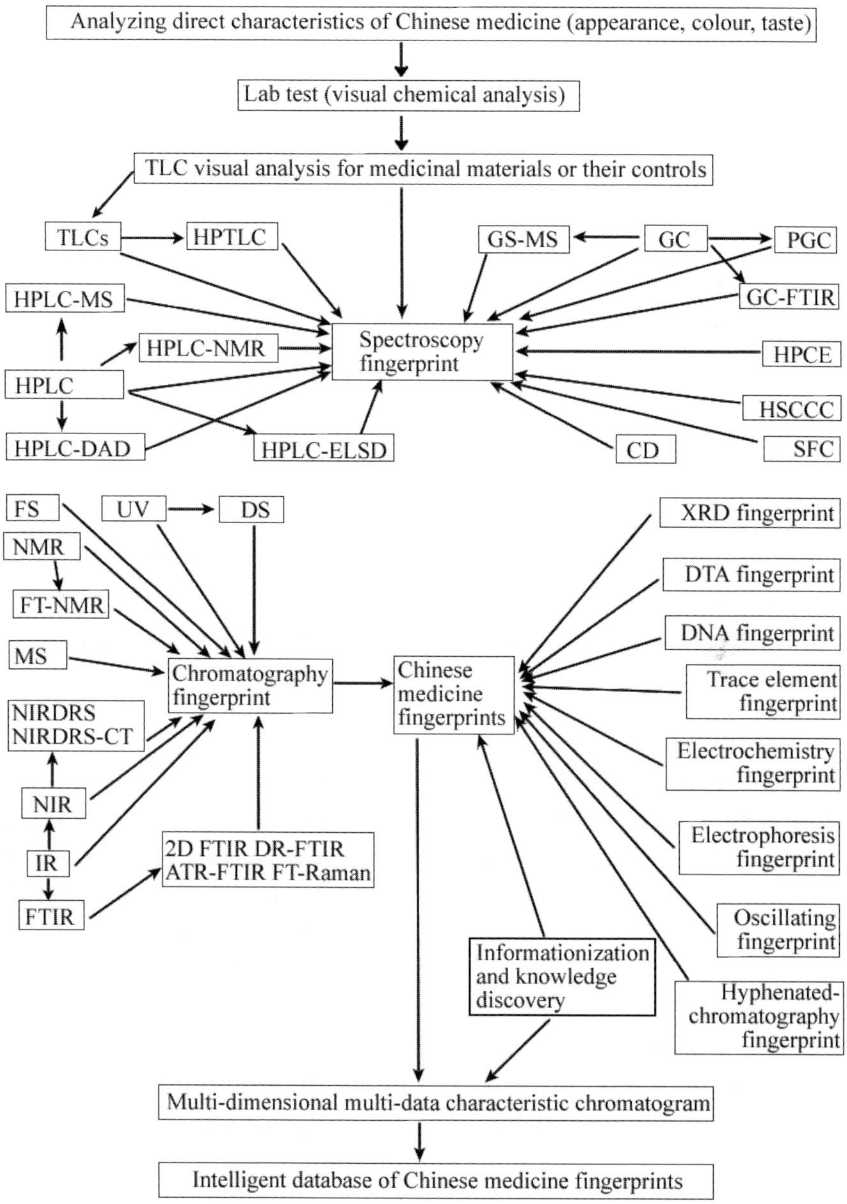

Fig. 1. Schematic diagram of the origin and development of Chinese medicine fingerprinting.

seng components have constantly been discovered. Currently, the known components include more than 30 types of saponins, 29 types of volatile oils, 15 types of amino acids, 29 types of minerals, 16 types of carbohydrates, 11 types of organic acids, esters, alkaloids, vitamins, sterols and many enzymes. A single medicinal material is so complex in nature, as it is, let alone the chemical components in a medicine formula. Based on this, it would be highly impractical if quality control were to be established on the basis of the study of every single chemical component. The requirement now is to study the entire species under the conditions where the chemical composition is not completely known. The fingerprints of species obtained through modern instrumental analysis, such as chromatography, spectroscopy and mass spectrometry analysis, present this possibility. Similarities in the fingerprints of several species obtained using the same instrument, the same operating procedures, and under the same test conditions reflect similarities in their attributes. Even though one may not understand the composition represented by each characteristic peak in the fingerprint, and may not have a complete knowledge of the chemical composition of the species, this does not influence the assessment of the uniformity of the species. It can provide qualitative analysis, as well as semi-quantitative analysis.

In fact, from the point of view of the development of analytical chemistry, analysis of pure substances is not particularly challenging. Yet many disciplines, such as environmental chemistry, have highlighted the requirement to carry out rapid qualitative and quantitative analysis of complex organic compounds. The trend towards the use of instrumentation for analysis and the more complex chemistry system are two important characteristics of modern analytical chemistry. At the same time, chemometrics has been developing rapidly. From the perspective of modern chemometrics, in the traditional method of data processing in instrumental analysis, such as for spectroscopy, only the peak value is taken; for chromatography, only the peak area is calculated. As a consequence, large amounts of useful information are wasted. Chemometrics, which uses statistics, applied mathematics and computers, is able to extract the maximum amount of qualitative and quantitative chemical information from samples with different characteristics. Research in this area also suggested the classification of the analysis system into white, grey and black multi-component systems according to the extent of prior knowledge of the qualitative composition. Modern analytical chemistry, especially the development of chemometrics, has set the technological foundation for the use of Chinese proprietary medicine fingerprinting for quantitative and qualitative analyses.

We have to control the intrinsic quality and at the same time ensure the efficacy of Chinese medicine. We have come to a historical moment in the development of analytical chemistry, where we are able to resolve the quantitative and qualitative analyses of complex multiple components. Hence, there is a need to make full use of the conditions to resolve the problems encountered, break the deadlock and establish a new era where qualitative and quantitative analyses of Chinese medicine are gradually made possible.

Fingerprinting as the Critical Technology for the Modernization of the Chinese Medicine Industry

The development of the Chinese medicine industry follows a modernization process. From improvements of plant areas, plant buildings, technologies and equipment, to innovations in packaging and dosage formulation, it is now time to modernize the quality standards of Chinese proprietary medicine. The modernization of the production of Chinese proprietary medicine will eventually be reflected in the modernization of the controllable quality of the products. The controllability refers to the fact that the quality standard of a product should ensure the stability of the drug efficacy. In addition, the quality of the products should be controllable during the entire manufacturing process. These two points are the most significant problems in the modern production of Chinese proprietary medicine. At present, the existing quality standard is not good enough to control the stability and efficacy of Chinese medicine products, and it is also difficult to ensure product consistency during the manufacturing process. All these problems come from our limited understanding of the different components in a product.

Quality control through fingerprinting is expected to significantly increase the level of controllability, and promote the modernization of a product and its production process. When fingerprinting is applied, the quality of the raw materials will need to be tested by the company and the storage management methods will need to be improved. To ensure consistency in the quality of the final products, the dosages of raw materials should be adjusted so as to make up for the differences in quality in different batches of materials. To ensure consistency of products, production processes should be studied to establish strict control parameters. To ensure consistency of operations and end results, management of the manufacturing processes and personnel must be improved. To meet final product specifications, the quality of intermediates should be monitored; technical strategies should be studied to ensure that material feeding in every step meets specifications. To ensure the consistency of equipment conditions and equipment usage, the equipment models should be carefully selected and equipment maintenance should be improved.

Since there are quality control requirements for fingerprints, we now have the criteria for evaluating the impact of technology and equipment changes on product quality during the manufacturing process. With the understanding that quality comes first, one can boldly experiment with various new technologies and equipment.

Fingerprinting technology is closely related to manufacturing procedures. There are many variable factors in this technology, which will cause inconsistencies in the fingerprints, yet the many adjustable factors broaden the field of research. It is similar to the difference between fluidized bed granulation and mechanical granulation, or that between spray drying and box drying. Fluidized bed granulation and spray drying present a lot of variables and are not easily mastered. But although expertise is required for these methods, if they are mastered by the operator, manipulations become very flexible. The industry of Chinese proprietary medicine also needs to overcome some technical and management obstacles when using fingerprinting technology. At the same time, a lot of technical secrets are hidden in the construction of the fingerprint of a new product. In the past, Chinese proprietary medicine formulae were trade secrets. Now, even if a formula is publicized, without the technical know-how to construct a qualified fingerprint, the formula is of no use. Fingerprints have become technical barriers to protect a company's new products. Unlike patent protection, this kind of barrier is not restricted by any time limit.

The use of fingerprinting technology for quality control will not only promote the application of new technologies and equipment, but will also open a new channel for research into new Chinese proprietary medicine. The effectiveness of various technologies can be compared by studying the different chromatograms obtained using the same medicine formula but with different production technologies, so that the most optimized technology can be selected.

The difficulty of fingerprinting technology will certainly drive the centralization and standardization of Chinese proprietary medicine production. It plays a guiding role in the modernization of Chinese proprietary medicine.

Fingerprint Quality Control as a Technical Link to Drive the Modernization of Chinese Medicine

Chinese agriculture has entered a new phase. While maintaining a stable increase in yield, it is paying more attention to increasing product quality, modifying and optimizing the plant species, and producing high-quality agricultural products. The government has announced guidelines to greatly increase the overall quality and efficiency of agriculture through science and technology. The government has also announced guidelines relating to the mode of production, saying that "industrialization should become an important approach to advance agricultural modernization, to encourage and support agricultural products processing and marketing companies to help farmers enter the market, and finally form a profit- and risk-sharing organization"[a].

It is also time for the agriculture of Chinese medicine to pay attention to quality and authenticity. How much attention will be paid to quality in the agriculture of Chinese medicine depends on how much attention will be paid to the quality of raw materials by the Chinese proprietary medicine companies. The Chinese medicine industry has been making great progress in the recent 30 years. The weakest link is

[a]The 10th Five-Year Plan for National Economic and Social Development. (2011, July 1). http://cpc.people.com.cn/GB/64162/71380/71382/71386/4837946.html

the purchase of raw materials. Some Chinese medicine companies do not pay attention to the quality of raw materials. They purchase whatever is cheapest, resulting in a situation where attention is paid to quantity, not quality, in the cultivation of Chinese medicine. Medicinal materials, decoction pieces and proprietary medicine are the three mainstays of the Chinese medicine industry. Among them, Chinese proprietary medicine has the highest market share. Seventy percent of the medicinal materials are sold to Chinese proprietary medicine companies. Being the main player in the Chinese medicine industry, the Chinese proprietary medicine industry now plays a leading role in Chinese medicine agriculture.

For a long time, many Chinese proprietary medicine companies were content with using raw materials that merely complied with quality requirements. This was partly a result of paying too much attention to profits, and partly because the quality standards of Chinese proprietary medicine lack specifications on the quality of raw materials. To apply fingerprinting technology in quality control, the quality of the raw materials must be stable. To ensure the quality of the final products, the companies of Chinese proprietary medicine must pay attention to the quality of medicinal raw materials and play a leading role in Chinese medicine agriculture. This kind of leadership role focuses on three aspects. The first aspect is the quality requirements of raw medicinal materials. Most farmers know that Chinese medicine companies need genuine and high-quality medicinal materials. The second aspect is to lead or guide the farmers towards industrialization. The Chinese proprietary medicine industry should pay more attention to helping farmers establish raw material bases and should consider the raw material bases as their first workshop. This is such that the quality of the final products is built upon the high quality and stability of the raw medicinal materials. The third aspect is to begin research into the cultivation techniques used for raw medicinal materials. This means bringing in agricultural technology, selecting better places of origin, and setting up the SOP for producing medicinal materials. In summary, the Chinese proprietary medicine industry should play a leading role in promoting Chinese medicine agriculture to achieve GAP. For major raw materials, companies should put effort into establishing bases that meet the GAP requirements.

The selection of places of origin of medicinal materials has great significance in base construction, development of SOP for medicinal material production, and application of GAP. There is much data demonstrating that there are huge differences in the concentrations of components in the same kind of medicinal materials if they have different origins. For many herbs, there could be a difference of more than 10 times in the main component concentrations when we compare herbs with different origins. For example, chlorogenic acid in honeysuckle ranges from 0.45% to 3.42%; icariin in *Herba Epimedii* ranges from 0.14% to 3.00%; astragaloside A in astragalus root ranges from 0.013% to 0.30%. That is why the selection of bases in the country should be based on both authenticity studies and scientific analysis.

Developing an SOP for the production of raw medicinal materials is very important to achieve standardized production and consistent quality. With an SOP, farmers will cultivate medicinal materials according to it, just like production workers manufacture medicinal products according to an SOP. In this case, GAP is similar to GMP. How to organize the farmers' efforts is a big issue. Some companies already have some successful experiences, such as the *Radix scutellariae* base in Zhangjiakou of Hebei Province setup by the Second Factory of Harbin Chinese Medicine; the *Radix salviae miltiorrhizae* base in Nanluo of Shanxi Province setup by Tianjin Tasly Inc. and the saffron crocus base in the suburbs of Shanghai setup by Shanghai Pharmaceuticals Pte. Ltd. There are various types of organization and it is important to experiment with different types of setup.

Besides setting up bases for cultivated species, research into converting wild species to cultivated species should be conducted actively according to industry needs. Bases should also be setup for successfully converted species. In brief, an industry cannot be established on the grounds of nearly exhausted resources. Building a medicinal material base will bring several advantages to the industry, enterprises and farmers, such as stable supply and demand, stable prices, stable production and stable quality. For the Chinese medicine industry, the industry plays a leading role in its modernization. A modernized industry will lead to a modernized agriculture. Only then can the sustainable and coordinated modernization of the Chinese medicine industry be ensured. One can say that without the modernization of Chinese medicine agriculture, the modernization of Chinese medicine is impossible. The use of fingerprinting has already pushed some Chinese medicine companies to seriously consider the issues of setting up bases for raw materials, the relationship between the Chinese medicine industry and Chinese medicine agriculture, and the relationship between Chinese medicine agriculture, carrying out national policies and achieving modernization. This shows how important this technology is in driving the Chinese medicine industry and Chinese medicine agriculture to jointly achieve modernization. This is only the beginning. The scaling up of the industry has brought about centralized and standardized (GAP) production of Chinese medicine agriculture, and its effect on Chinese medicine agriculture will become more and more significant.

Fingerprinting as an Essential Quality Control Technique

In the modern world, "safe, effective, and controllable" are internationally recognized requirements of medicaments. There is also a growing common understanding in the quality control methods in Chinese proprietary medicine in recent years. The major Japanese companies producing Chinese medicine have been using HPLC fingerprinting for quality control since the 1980s. They produced standard fingerprints using decoctions of medicinal teas prepared according to prescription formulae. The raw materials, formulae and operating procedures used for mass production were strictly controlled to ensure that the final product fingerprints would be consistent with the standard fingerprints. Europe has always valued the therapeutic effects of herbal medicine. The Europeans have also been using fingerprinting to control the quality of herbal medicine. One typical example is the extraction of *Folium Ginkgo* jointly developed by researchers in Germany and France. During their research, they found that the therapeutic effect of *Folium Ginkgo* extract comes from the synergistic effects of the substances in the extract. They conducted a long-term research study and found that isolated lactones or flavones from the extract do not have the same effect as that of the extract as a whole. This finding is consistent with our understanding of the effects of Chinese proprietary medicine. HPLC fingerprinting should be used for the quality control of this kind of "whole materials." For example, *Hypericum perforatum* has been used in Germany to treat depression. It was introduced to the United States and was very popular in the US herbal market in 1998. In the US, the content of hypericin was used as a quality indicator. However, some studies in recent years have discovered that this component does not have antidepressant effects. Hence, the fingerprinting technology that has been used for quality control by German companies is regarded as a more reliable method. In recent years, the US FDA has listed HPLC fingerprinting as the quality control method for mixtures of substances in its botanical drugs guidelines. In brief, fingerprinting is internationally recognized as the quality control method for Chinese proprietary medicine and herbal extracts. Therefore, a quality control system using fingerprinting technology should be established according to the unique features of Chinese medicine.

Requirements and Construction of Fingerprints

Requirements of Fingerprints

Chinese medicine fingerprints should satisfy the practical requirements of specificity, reproducibility and practicability.

Regardless of whether it is a Chinese medicinal material, Chinese proprietary medicine or Chinese medicine formula, the fingerprint must be able to display the characteristic, also known as the specificity or uniqueness, of that Chinese medicine (herb or formula). Of course, there is the possibility that one fingerprint is not adequate to represent the full characteristics of a Chinese medicine, and so there may be a need to use several fingerprints to demonstrate the different characteristic profiles of a certain Chinese medicine (herb or formula), in order to form the full picture. But for each fingerprint there should still be the requirement of specificity.

Fingerprinting is mainly used to demonstrate and control the chemical constituents of Chinese medicine as a whole. Thus there is a need for good reproducibility, i.e. for the same sample, under the same operating conditions, the level of reproducibility of the result must be good. Hence, according to the differing requirements, one should select the most suitable analytical method for fingerprint construction. One of the main characteristics of Chinese medicine is its complexity. It involves uncertainty and some fuzziness. These problems affect the reproducibility, and could be eliminated in the data-processing steps (such as in the assessment of degree of similarity) through the use of suitable methods.

The practicability of a fingerprint refers to the choice of different analytical methods according to different applications to achieve the different requirements. For example, for quality control, one should consider the use of conventional facilities or equipment found in the manufacturing enterprise or drug analysis department to setup the corresponding method. Usually, these include photometers, GC and HPLC. In the case of using fingerprinting to study the theory of herb compatibility (配伍) or the development of new medicine, especially those in areas related to chemical composition, pharmacology and drug efficacy, one should consider the use of hyphenated technology such as GC-MS and HPLC-DAD-MS/MS to obtain a large amount of data. This is preferable due to the acquisition of unambiguous results.

Methods of Constructing Fingerprints

With the rapid development of contemporary chromatographic analysis and photometric analysis, more and more advanced techniques are applied to the quality analysis of Chinese medicine. These are powerful tools for the construction of the Chinese medicine fingerprints. The commonly used methods of constructing fingerprints are the photometric method, chromatographic method, X-ray diffraction method, thermal analysis method, electrophoresis, hyphenated chromatographic technology, molecular biology technology, electrochemistry method, scanning electron microscope, computer image analysis and chemical oscillation technology. In actual applications, we can select one or more methods according to our needs.

Photometry

Photometry refers to the use of light of a certain wavelength to scan the sample and obtain the characteristic spectra and data. The main methods include ultraviolet (UV) spectrum, infrared (IR) spectrum, fluorescence spectrum (FS), nuclear magnetic resonance (NMR) and mass spectrometry (MS). In addition, there is atomic absorption spectrometry (AAS), inductively coupled plasma (ICP) and X-ray fluorescence (XRF).

UV spectrum

UV spectrum is the absorption spectrum that demonstrates the absorption of energy in the UV and visible light regions by electrons in the molecule to produce electronic transitions. In Chinese medicines, some of the constituents possess unsaturated or conjugated double bond structures, which under UV light (200–400 nm) can undergo changes in the atomic or electronic active states of the chemical substances. A portion of the energy will be consumed while the rest is transmitted, passed through a prism or diffraction grating to be separated into a non-continuous spectrum based on the wavelength. Dissimilar substances produce dissimilar UV absorption spectra. Since different Chinese medicines exhibit diversity in the unsaturated components, their UV spectra would also demonstrate characteristic differences (maximal absorption wavelength, minimal absorption wavelength, peak shoulder, absorption coefficient, absorbance ratio, number of absorbance peaks, peak shape, peak height, etc). The UV absorption spectrum of a Chinese medicinal herb is a complex spectrum formed by the overlap of the characteristic absorption spectrum of multiple constituents. Under certain conditions, the complex combination of the many constituents of a Chinese medicine possesses a certain regular pattern, which shows up in the UV spectrum with a definite specificity and stability. For Chinese medicines that are hard to distinguish or have close kinship, one can select the multiple-solvent UV spectrum method or derivative spectrum method. The derivative spectrum method is able to eliminate some interference in the original spectrum which is unrelated to the absorbance. It can also resolve the problem of overlapping absorption peaks in the UV spectrum. With the increase in the number of derivatives, the absorption band becomes narrower, the peak shape becomes sharper, resulting in an increase in the spectral resolution.

IR spectrum

Infrared spectrum is an absorption spectrum that reflects the change in the vibrational and rotational energy levels of the intermolecular bondings. IR spectrum possesses high specificity. Every single compound has its characteristic IR spectrum. A Chinese medicine is a mixture of many chemical constituents. Its IR spectrum is the overlap of the absorption in the infrared region (400–4000 cm^{-1}) by the overall functional groups of each constituent in the mixture. Under the same conditions, the chemical composition in the same Chinese medicine would have a relatively constant quality and quantity, which is distinguishable from other Chinese medicines. Hence the overlapped IR spectrum is stable and specific. Spectrum analysis focuses mainly on the comparison of the characteristic outline (graphical shape of the entire spectrum) of the IR spectrum as the main absorption peaks do not need to be analyzed. Within the scanned wavenumber range, it is sufficient to compare the wavenumbers of the absorption peaks, the peak shape and peak intensity of the same wavenumber, and the differences in the appearance of the "fingerprint region." With the fast development of instrumental analysis and chemometric and computational techniques, for a complex mixture system like Chinese medicine, many more new methods and new technologies which are even more practical have been developed. Examples are the Fourier transform IR (FTIR), two-dimensional IR correlation spectrometry (2D FTIR), near IR (NIR), Fourier transform–Raman (FT-Raman), attenuated total reflectance FTIR (ATR-FTIR), diffuse reflectance FTIR (DR-FTIR), near-infrared diffuse reflectance spectroscopy

(NIRDRS) and near-infrared diffuse reflectance spectroscopy visualization convolution transform (NIRDRS-CT), which are even more suitable for fingerprint research and construction.

Fluorescence spectrum

Certain components in Chinese medicines, usually molecules with conjugated double bonds or aromatic systems, under UV light, are able to absorb light energy of a certain wavelength, and emit light of an even longer wavelength. When the UV light stops, the photoluminescence instantly disappears. This light is usually luminous visible light, which is called fluorescence. Fluorescence spectrum includes excitation spectrum and emission spectrum. Excitation spectrum refers to the relative efficiency in using different wavelengths of excitation light to cause a substance to emit a single wavelength of fluorescent light. This means that the fluorescent emission monochromator is fixed, while the excitation monochromator scans the wavelength to record the corresponding fluorescence intensity at various wavelengths. Emission spectrum refers to the relative efficiency in using a single wavelength of excitation light to cause a substance to emit a different wavelength of fluorescent light. This means that the excitation monochromator is fixed, while the emission monochromator scans the wavelength of the fluorescent light produced by the substance, to record the fluorescence intensity at various wavelengths. Different molecular structures of a substance absorb UV light and emit fluorescent light of different wavelengths. Its fluorescence spectrum has a definite specificity; the highest excitation wavelength (λ_{ex}), highest emission wavelength (λ_{em}) and peak number are characteristic constants used for analysis. Sometimes, to improve the sensitivity and selectivity of the analytical method, weakly fluorescent substances are reacted with some fluorescent reagent to obtain strongly fluorescent products. Despite the small number of fluorescent peaks, one is able to obtain both the excitation and emission spectra of the substance, which provides more information compared to the UV and IR spectra, and is more effective for identification.

Nuclear magnetic resonance

Nuclear magnetic resonance reflects the changes in the energy level of the magnetic atomic nuclei (H atom in H spectra, C atom in C spectra) in a molecule when under the influence of a strong magnetic field and radiofrequency, which produce nuclear magnetic resonance. It is one of the important methods for the identification of an organic compound, providing structural information about the compound, such as each proton's chemical shift, number and coupling. In Chinese medicine (characteristic whole extraction), under the effect of electromagnetic radiation with a certain frequency, very long wavelength (10 cm to 100 m) and very low energy level, atoms of certain elements (usually H) in the chemical composition are able to absorb electromagnetic radiation to produce ^{1}H-NMR spectrum of peak intensity versus absorption frequency. The composition of Chinese medicine is complex. If a certain procedure is used to obtain the total extract of the characteristic chemical component(s), and at the same time, the contents of these characteristic chemical components are relatively constant, then, under the conditions of standardized extraction, a strict correlation would exist between the ^{1}H-NMR spectrum of a plant-based Chinese medicine and the plant strain. Some research studies have shown that the ^{1}H-NMR fingerprints of Chinese medicines possess a high degree of specificity and reproducibility by carrying out an analysis of the characteristic resonance signal and data (δ_{ppm}) of the ^{1}H-NMR.

Mass spectrum

Mass spectrum is a spectrum formed from the line-up of the mass to charge ratio (m/z) of the charged particles (ions). It is the mass spectrum of physical particles that uses a high-speed electron stream to bombard sample molecules, causing them to break down into many ion fragments. Then, in the magnetic field, amid an alternating electric field or vacuum environment, these fast-moving ion fragments are separated. Based on these ion fragments, conclusions are made about the molecular weight and the molecular structure. Under the bombardment of electrons, characteristic extracts of Chinese medicines are broken down into fragments in the mass spectrum to produce the EI-MS spectrum of the chemical component in the extract. Since different Chinese medicines contain different components, the molecular ion base peak and the fragment peaks (m/z) shown in the mass spectrum are also not the same. Hence mass spectrometry possesses stronger fingerprint characteristics.

Chromatography

Chromatography is the fastest developing and the most widely used modern analytical method in the research area of Chinese medicine. The principle of chromatography is based on the different interphase distribution of substances and the results of their separation. In the recent 100 years of development, various types of chromatographic techniques have been created. Chromatographic techniques have extremely strong separating power and adaptability, and are the top choice in Chinese medicine fingerprinting. The commonly used methods are thin-layer chromatography (TLC), gas chromatography (GC), higher-performance liquid chromatography (HPLC), high-performance capillary electrophoresis (HPCE), high-speed counter-current chromatography (HSCCC), supercritical fluid chromatography (SFC) and circular dichroism (CD).

Thin-layer chromatography

Thin-layer chromatography is currently the most widely used method in fingerprinting, as it is fast, economical, reliable, easy to operate, and supports a wide range of applications with good reproducibility. The unique strength of TLC is that it provides a visual image of visible light or fluorescent chromatograms. It can be further combined with a chromatographic scan or digital processing to obtain a chromatogram that shows the different graduation outlines and the corresponding integral data. It is especially suitable for daily analysis and field inspection, and is the main method for Chinese medicine identification. Modern TLC (that is, 2-D chromatography), through the help of advanced technology and computation technology, has developed to the stage where instrumentation, automation, computation and hyphenation of other chromatographic techniques are used. Special technologies begin to appear, such as high-performance TLC (HPTLC), thin-layer chromatography scan (TLCS), thermal micro application separation (TAS) TLC, thin-layer chromatography–gas chromatography (TLC-GC) and thin-layer chromatography–infrared spectrometry (TLC-IR). Of these, the scan chromatogram obtained by using a fixed wavelength to scan the spots in TLCS is much more objective and accurate than the visual estimation of TLC. On top of that, the peak area of each peak and standard peak ratio can be quantified, which has a greater significance for fingerprint analysis.

Gas chromatography

Gas chromatography allows high resolution and fast analysis. The chromatogram obtained has good reproducibility and good resolution. Since it is a type of closed-system chromatography, there are few influencing external factors, the stability is good, and there is a wide choice of detectors. It is especially suitable for research in Chinese medicine that contains volatile components. In recent years, with the advancement in technology, research and practical needs, new methods and new technologies with even stronger practicalities have appeared. Examples are capillary gas chromatography (CGC), gas chromatography retention index spectrum (GCRIS), pyrolysis–gas chromatography (PGC), head-space gas chromatography (HSGC), gas chromatography–mass spectrometry (GC-MS) and gas chromatography–Fourier transform infrared spectrometry (GC-FTIR). Of these, the most prominent one is pyrolysis-gas chromatography. It is easy to operate, its samples do not require pre-treatment, and it provides a large amount of data. It also has a wide range of applications. For certain Chinese medicines that cannot be directly analyzed with GC, PGC can be used. PGC combines pyrolysis and gas chromatography technology. The sample is put into a pyrolysis device and heated under certain conditions to pyrolyze it into volatile small molecules instantaneously. The molecular products are transferred into the GC column by a carrier gas. After separation, a reproducible pyrolysis–gas chromatogram which reveals the fingerprint characteristics will be obtained from the recording instrument. GCRIS is based on the GC retention indices of the volatile components in Chinese medicine. From these, the retention indices of certain characteristic components are selected to make up the gas chromatographic retention index spectra, such that the characteristic fingerprint is more distinctive. GC-MS is widely applied and is able to provide "on-line" information about the chemical structure of the main components in the fingerprint. Fast and sensitive, it is a powerful support tool for fingerprinting. The use of supercritical CO_2 fluid extraction as pre-treatment for GC-MS has also tremendously broadened the scope for Chinese medicine fingerprint construction.

High-performance liquid chromatography

High-performance liquid chromatography has the characteristics of good separation efficiency, fast speed of analysis, good sensitivity, good stability and reproducibility, wide selection of mobile phase and detectors, and re-usable column, making it extremely suitable for the construction of Chinese medicine fingerprints. HPLC is not limited by the sample volatility and heat stability and is widely used for non-volatile components. With the popularity of HPLC, its applications are increasingly expanding. Currently, it is the main and most commonly used method for the construction of Chinese medicine fingerprints. With the development of new, high-efficiency common detectors, such as the introduction of diode array detectors (DAD), evaporative light-scattering detectors (ELSD), electrospray ionization–mass spectrometry (ESI-MS), MS/MS, the applicability and resolution of HPLC have increased significantly, making possible the analysis of complex samples. From HPLC-DAD, three-dimensional HPLC chromatograms can be obtained. With the peak retention time and spectral data integrated into a single chromatogram, it is suitable for test samples with complex components that absorb light in the UV region. HPLC-ELSD is able to analyze substances that do not absorb UV light. Through HPLC-UV-MS, the HPLC-UV and HPLC-MS fingerprints can be obtained and the main peaks can be classified. The total ion chromatogram obtained from HPLC-ESI-MS contains even more data, and possesses greater specificity and fingerprint characteristics. HPLC-DAD-MS/MS can be used to establish multi-dimensional fingerprints and total ion chromatograms at the same time.

High-performance capillary electrophoresis

High-performance capillary electrophoresis is a new separation and analytical technique that has been rapidly developing in the last 10 to 20 years. It is a liquid separation technique that uses a high-voltage electrical field as the driving force, the capillary as the separation channel, and separates components of a sample based on their differential mobility and separation behavior. It has the characteristics of high separation efficiency and fast analytical speed. It has many types of separation models, is easy to clean and maintain, and is able to analyze aqueous solutions directly. It is increasingly showing prominent prospects in its applications in the research of Chinese medicine fingerprinting. HPCE is suitable for the analysis of most chemical components, especially the separation of large biomolecules such as peptides and proteins found in animal-based Chinese medicines. The chemical composition of Chinese medicines is complex and diversified; the molecules are of different sizes. HPCE is able to analyze both large molecules and small molecules (such as proteins and phenolic acids) simultaneously. The fingerprint thus obtained has a greater ability to reflect the characteristic components of the Chinese medicine. A variety of separation models are available; the ones that have been developed are capillary zone electrophoresis (CZE), capillary gel electrophoresis (CGE), micellar electrokinetic capillary chromatography (MECC), capillary isotachophoresis (CITP), capillary isoelectric focusing (CIEF) and capillary electrochromatography (CEC). Of these, CZE and MECC are the two most commonly used models in the research of Chinese medicine fingerprinting. The reproducibility of HPCE can be improved through mobility comparison, reference correction and retention index methods.

High-speed counter-current chromatography

High-speed counter-current chromatography is an internationally popular liquid-liquid separation technology. The technology employs the principle of dynamic liquid-liquid separation, making use of two non-miscible solvents moving relative to each other. The two phases which are in dynamic equilibrium will separate the sample components based on their distribution ratio. It has the strengths of high separation efficiency, large overloading capacity, no irreversible adsorption on solid carrier and less solvent consumption. HSCCC technology possesses very good adaptability; since the combinations and ratios of solvent systems are infinite, theoretically it can be used in the separation of samples of any polarity, which is an exceptional strength in the separation of natural products. Its resolution and reproducibility are quite good, and the operation is simple and easy to handle. There is little requirement for sample pre-treatment; good separation effect is achieved with simple extraction or even without pre-treatment. Injections can also be repeated in gradient mode or reversed-phase mode. This technology has a good rate of recovery since HSCCC does not need a solid-state carrier, and therefore the problem of adsorption and degeneration does not exist. So long as the separation conditions are properly adjusted, a good recovery rate can be expected. This is why HSCCC is very suitable for the separation and analysis of Chinese medicine components. Many research studies have shown that HSCCC is able to separate components that are not separable through HPLC. Research on the use of HSCCC in Chinese herbal medicines is still in the initial phase. Currently, it is mainly used in the separation analysis of the active components in Chinese herbal medicines. For the past few years, with the advancement of instruments and methods, the combination of HSCCC and mass spectrometry (HSCCC-MS) has successfully been applied in the separation, analysis and identification of chemical components in Chinese medicine. In the research of Chinese medicine quality control and analysis, especially in the field of fingerprint analysis, HSCCC has broad and favorable application prospects.

Supercritical fluid chromatography

Supercritical fluid chromatography utilizes a supercritical fluid as the mobile phase. Through the use of pressure control, the density of the mobile phase is adjusted to effect the change in the solubility of the substances to be separated, such that the various substances go through dissolution and resolution separately. The supercritical fluid is a powerful solvent, with good fluidity and a fast rate of mass transfer. It combines the strengths of HPLC and GC and overcomes some of their weaknesses. SFC has an even wider scope of application than GC, gives better quantitative results than HPLC, requires a shorter run time and easier sample pre-treatment, and is more readily hyphenated with large analytical instruments, such as mass spectrometry, FTIR and NMR. SFC has a very wide range of compatibility. Theoretically, SFC can be used for any compound that can be separated and analyzed by HPLC. Even for substances that cannot be separated by HPLC, SFC can be used to achieve good separation.

X-ray diffraction

X-ray diffraction is the main method used to study the phases and crystal structures of substances. When a certain substance undergoes diffraction analysis, the substance produces different degrees of diffraction phenomenon, under the effect of X-ray. Its composition, crystal structure, type of molecular bonding and molecular geometry determine the distinctive diffraction pattern of that substance. If the material is a mixture, the diffraction pattern is obtained from the superimposed diffraction effect of each component. As long as the composition of the mixture is constant, the diffraction pattern can be characteristic of the mixture. Though the composition of Chinese medicines (including organic and inorganic multi-phase systems) is complex, when the influencing factors such as the site of cultivation or timing of collection are relatively constant, its composition and hence diffraction pattern will be constant as well. Since the compositions of Chinese medicines are all different, their diffraction patterns have distinctive characteristics. X-ray diffraction fingerprinting has strong fingerprint characteristics, provides immediate information about the sample composition and gives a stable, reliable pattern. XRD can be classified into the single-crystal X-ray diffraction method and powder X-ray diffraction method. Powder X-ray diffraction is fast and simple, provides a large amount of pattern information, possesses strong fingerprint specificity and multiple evaluation indices (such as diffraction pattern, *d* value and relative intensity), and is stable and reliable. It is suited for the fingerprint construction of Chinese medicines. The powder X-ray diffraction–Fourier analysis method is based on converting a diffraction signal through Fourier transformation to obtain the X-ray diffraction pattern and the topology pattern, so as to construct a simpler X-ray diffraction Fourier fingerprint that can reflect the characteristics of the overall structure of a Chinese medicine.

Thermal analysis

During the process of heating or cooling, many substances will undergo physical or chemical changes such as dissolution, solidification, decomposition, reaction, adsorption, desorption and crystal transformation, with the release or absorption of heat. The technology that studies and analyzes such changes is known as thermal analysis. According to the type of analysis, it can be classified into thermogravimetry (TG), differential thermal analysis (DTA) and differential scanning calorimetry (DSC). Of these, DTA is the most commonly used. This method studies the temperature profiles of a sample and a reference material against time or temperature, when both are heated at the same rate under the same conditions. The result of the analysis is shown in a thermogram. Different Chinese medicines will have their own characteristic thermograms.

Electrophoresis

Electrophoresis is the directional movement of charged particles in a buffer under the influence of an electric field. Charged components in

Chinese medicines, such as organic acids, proteins, peptides, amino acids, alkaloids and enzymes, in an electric field of a certain strength and the same period of time, will migrate in different directions (towards the anode or cathode), at different speeds and different distances, according to the nature of the charge (positive or negative), number of charges and the molecular weight of each component. This gives rise to a characteristic band number, band distribution and band colors on the electrophoresis map. Based on the different matrices used, conventional electrophoresis can be classified into paper electrophoresis (PE), polyacrylamide gel electrophoresis (PAGE), cellulose acetate membrane electrophoresis (CAME), etc. Of these, the PAGE method is more commonly used. This method only requires a simple laboratory setup, has high specificity and good sensitivity. High-performance capillary electrophoresis is a tremendous improvement on conventional electrophoresis (refer to section on chromatography). Conventional electrophoresis has an edge in the analysis of Chinese medicine that contains different compositions of peptides and proteins, but its results are more easily influenced by the test conditions. Hence, one should strictly control the consistency of the test conditions.

Hyphenated chromatographic technology

Chinese medicine prescription preparations have complex compositions. Very often the use of a single chromatographic method or condition is not able to generate a comprehensive fingerprint, and hence the intrinsic quality of the Chinese medicine is not fully and accurately reflected. Use of hyphenated chromatographic technology generates multi-dimensional and multi-data characteristic fingerprints, which are able to resolve the difficulty of demonstrating the complexity of prescription preparation in its entirety. Multi-dimensionality refers to the link up of several analytical instruments to generate the fingerprints. Each fingerprint provides complementary data, such that a clearer and more complete understanding of the complex sample is formed. Currently the most commonly used method is high-performance liquid chromatography or capillary electrophoresis–diode array detector–mass spectrometry–mass spectrometry to obtain multi-dimensional fingerprints. It includes the chromatograms from HPLC or CE (with the retention time of each component), online UV spectrum from diode array detectors, stage 1 mass spectrum (with the mass of each component) and stage 2 mass spectrum (with the characteristic fragment of certain components). Multi-data refers to the characteristic fingerprints which include data in the areas of chemistry and drug efficacy. Chemical data relates to the multi-dimensional fingerprints mentioned above. Drug efficacy data is the relationship between the chemical composition and drug efficacy. Through the determination of the mass ratio of each effective group (chemical groups that are confirmed by drug efficacy experiments) and its active components (each type of compounds), the dose-response relationship of the drug can be deduced from the change in the mass ratio of the chemical components. The construction of multi-dimensional and multi-data characteristic fingerprints not only resolves the difficulty of ensuring the efficacy and quality of Chinese medicine prescription in a more systematic and comprehensive way, it also provides a new solution to the problem of a lack of reference standards for Chinese medicine research. With the rapid development of HPLC-NMR, without the acquisition of pure standards, the identification of the structure of each chemical component in a mixture can be achieved through the use of several hyphenated chromatographic technologies.

Molecular biology technology

In recent years, people have been realizing that the nucleotide sequence contains hereditary information required for the building of organic compounds, and for the maintenance and reproduction of life. Different organisms contain different DNA sequences. The same organisms contain the same DNA sequence to maintain the same hereditary traits, and yet possess polymorphism such that every individual of the same species will have many differences in its body. Since the sequence of such nucleotides is relatively stable and different in each individual, modern molecular biology technology can be used to generate DNA fingerprints. DNA fingerprints are highly specific to the individual. Research methods on the construction of fingerprints using DNA molecules from the polymerase chain reaction (PCR) as markers can be classified into two categories based on the subject of analysis. One is the determination of the polymorphism of DNA genomes, such as random amplification of polymorphic DNA (RAPD) and arbitrary primer PCR (AP-PCR); another is the determination of the polymorphism of DNA fragments, such as restriction fragment length polymorphism (RFLP) and direct DNA sequencing. RAPD and AP-PCR markers are able to determine the polymorphism of DNA when the characteristic DNA sequence is not available, which is more suitable for the construction of DNA fingerprints. RAPD has the advantages of being fast, simple and sensitive. It has been widely applied in recent years and its application prospects are extremely positive. RFLP is the most commonly used method to construct DNA fingerprints. It requires no significant decomposition of the DNA. As this method is time- and effort-consuming, there are limitations to its use. Direct DNA sequencing is, through PCR, the use of a universal primer to amplify a specific fragment of heavily degraded DNA from the medicinal material. The fragments are then sequenced. This method has relatively high repeatability and stability. It is able to minimize errors due to the characteristics of the medicinal material and yet has high specificity.

Each of the methods or technologies mentioned above has its adaptability and specialty in its applications. But there will be some problems when they are used in the analysis of the quality of Chinese medicine. Hence, in the research on the construction of Chinese medicine fingerprints, the choice of method should be based on the actual conditions and the actual analysis. In the case of mixtures, the specificity of UV and IR for identification and resolution is poor; the data processing requires the use of computerized pattern recognition technology or fuzzy mathematics. NMR and MS experiments are very costly; hence their applications are not widely promoted. While the characterization of Chinese medicines using the spectrometric fingerprinting method can be used to differentiate them, unlike chromatographic fingerprinting, it is not able to demonstrate the overall nature of Chinese medicines, which are complex mixtures with different chemical components of various concentrations. When there are changes in the concentrations of certain components in the complex mixtures of Chinese medicines, due to limitations in sensitivity and selectivity, they are not clearly and accurately reflected in the spectrum. Hence, it is challenging to employ this method in the evaluation of the stability of Chinese medicine quality. While TLC is simple and easy to use, with a high rate of utilization, its main shortcoming is the low "column efficiency" and limited information provided. It is difficult to use TLC to demonstrate a complex system made up of up to hundreds of chemical components. Its sensitivity is also inadequate for distinguishing characteristic fingerprints that are distinguishable only through tiny details. Second, the external environment has a greater influence on an open system such as TLC, hence the standards for external environmental conditions (temperature and moisture) are quite high, and operations should be skilled and standardized. GC is suited for volatile chemical components. HPCE has the advantage of good resolution and is more suited for large molecular compounds, but its reproducibility has to be improved. HPLC fingerprinting is able to adequately resolve each complex component, demonstrate the overall chemical characteristics and express the mass ratio of each component. Multiple data can be obtained from HPLC online analysis. For samples with complex components, with fingerprint characteristics that are distinguishable only through tiny details, it is an effective analytical technique. Hyphenated technology is the most effective method of generating fingerprints, as GC-MS, HPLC-MS, HPLC-MS/MS, etc, are able to provide a large amount of various data, and they fulfill the requirements of analyzing the complex systems of Chinese medicines. However, the instruments are costly and their use is not easy to popularize.

In short, the more prominent or promising methods in Chinese medicine fingerprint research are HPLC, HPCE, PGC, SFC, HSCCC and hyphenated chromatographic technology. Currently, the best method for constructing Chinese medicine fingerprints is HPLC.

Data Processing of Fingerprints

Direct visual analysis and comparison

Based on the fingerprint appearance and the data obtained, direct visual analysis, comparison and evaluation of the characteristics of the test samples and reference fingerprints are carried out. This method is suited for qualitative, rapid identification or is used when there is little data in the fingerprint. Currently, this is still a widely applied method.

Quantifiable data comparison

Quantifiable data comparison refers to the introduction of a relative index, rate of overlap, eight strong peaks, *N* strong peaks, apparent abundance, etc, which are quantifiable data for the analysis, comparison and evaluation of test samples and reference fingerprints. This method is more objective, accurate and has its practicality.

Chemical pattern recognition technology

Chemical pattern recognition refers to the use of computers to classify or describe a substance based on its chemical composition. The fingerprints of Chinese medicines are relatively complex, and visual comparison is bound to affect the accuracy of the result. If it is combined with statistics, computerized fingerprint analysis and pattern recognition, then its application will be more accurate, reliable, fast and convenient. The main methods include principal component analysis (PCA), simple classification algorithm (SIMCA), nonlinear mapping (NLM), constellation graphing technique, fuzzy information analysis, grey relational grade cluster, hierarchical clustering analysis, fuzzy clustering analysis and artificial neural network (ANN). Chemical pattern recognition technology has been widely utilized in Chinese medicine fingerprint research.

Chemometric methods

Commonly used chemometric methods are quantitative structure-activity relationship, molecular modeling and optimization, and use of resolution (R_s), sum of relative peak area (SRPA), WSK which is related to resolution and sum of relative peak area, mutual information and other target parameters that are used as the quantitative evaluation standards for fingerprints.

Computerized software analysis

Based on chemical pattern recognition technology, computerized analysis software, such as the comparison analysis software that comes with infrared spectroscopy, was developed for the ease of study and application. Computer-aided Chinese medicine fingerprint similarity evaluation software has been used in the research and actual work of Chinese medicine fingerprinting.

Fingerprint Setup Procedure

The main steps of fingerprint setup are sample collection, method development, data analysis, sample evaluation and method validation. Sample collection involves collecting sufficient specimens that reflect the sample quality, to ensure that the test sample is representative and uniform. Method development refers to the choice of a suitable method to setup and study the fingerprints. Data analysis refers to the processing of the data obtained, the establishment of the commonalities and differences, and the evaluation criteria. Sample evaluation uses an established method to analyze many more unknown samples over a long period of time, in order to further study the feasibility and practicality of the method.

The Chinese medicinal material fingerprint setup procedure is as shown in Fig. 2.

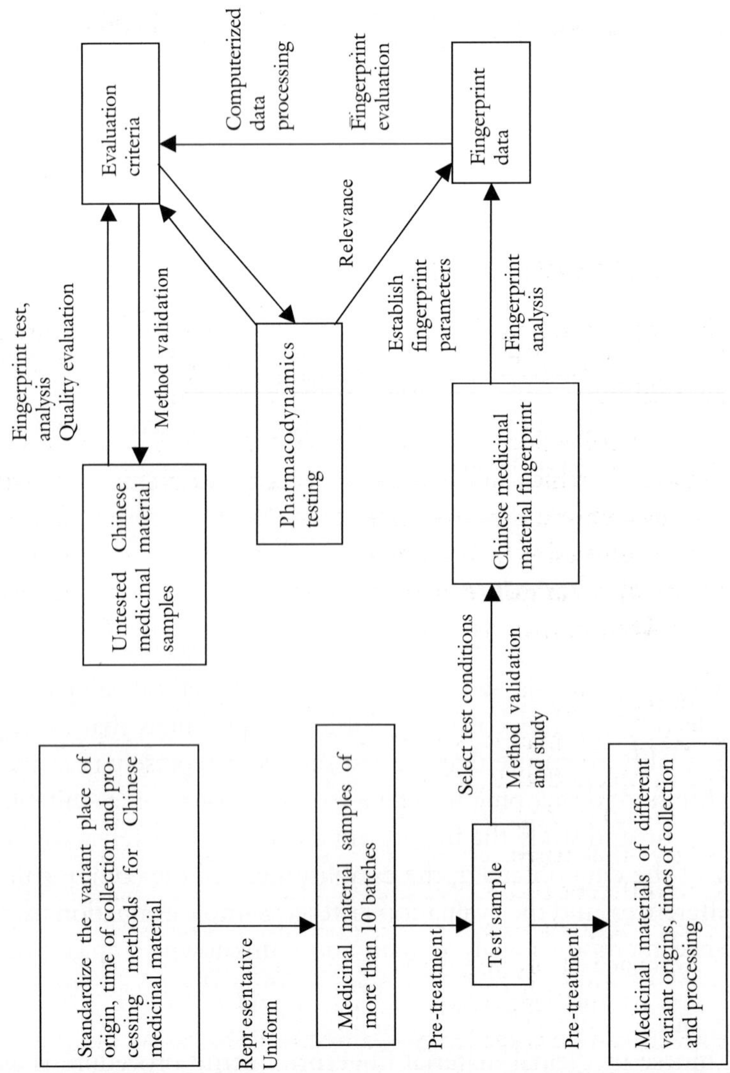

Fig. 2. Schematic diagram of Chinese medicinal material fingerprint setup procedure.

1.2

Research Methods and Technical Requirements of HPLC Fingerprinting

Sample Collection

Sample collection is the first and the most essential step in the research of HPLC fingerprinting. Since it is impossible to test all samples of the medicinal material and the growth environment, cultivation conditions, season of collection, processing and storage procedures of the medicinal materials which affect their secondary metabolites (active ingredients), the samples collected must be sufficiently representative. Considering the diversity of biological samples, only when an adequate quantity of a sample is used can we clearly demonstrate its characteristics. Hence, there is a need to collect more than 10 batches of samples.

The meaning of "batches" for drug samples is completely different from its meaning for manufacturing products. It refers to mutually independent samples, that is, samples from the same location, same channel or same collection time should not be considered as separate sample batches, so as to ensure that the test results are representative. Samples from different places of origin or from the same place of origin but of different grades may be viewed as different batches. Samples from the same place of origin may be different if they come from different plantations and land plots. All such samples that come from different sources can be considered as different batches. As the collection of medicinal material samples is limited by subjective and objective conditions, using more samples will lead to greater biostatistical significance.

The basic requirements for sample collection are as follows:

1. First, the medicinal material samples need to go through variant identification. Medicinal materials and their sliced or processed products should conform to national pharmacopeia, Ministry of Health standards, local standards or medicinal material processing regulations. The samples selected should be authentic, good-quality Chinese medicinal materials, or are Chinese medicinal materials produced in places of cultivation that conform to standard agricultural practices.
2. The variant, medicinal part, place of origin, time of collection, preparation and processing methods should be standardized for all medicinal material samples of plant or animal origin. The place of origin and the preparation and processing method should be standardized for mineral medicinal material samples. For Chinese medicinal materials that have multiple variants or multiple medicinal parts, a particular variant or a particular medicinal part should be selected. At the same time, one should carry out the analysis and comparison of the fingerprints of representative samples of the different variants and medicinal parts. This is to determine the common characteristics and specific characteristics of the fingerprints of each variant and each medicinal part. The common characteristics of the fingerprints are used to determine the quality of that particular medicinal material; the specific characteristics are used to determine the type of variant and medicinal part of the medicinal material, and also to determine the quality of that particular variant or medicinal part. Usually, a particular place of origin or a particular season of collection is selected. If multiple places of origin or multiple seasons of collection are chosen, one should inspect the fingerprints of representative samples from different origins or different seasons of collection to look for uniformity.
3. To ensure the authenticity of the samples, there should be a complete and original record of the sample collection, the content of which includes

(a) Name of medicinal material: Medicinal material name and its origin (the scientific name of the original plant or animal, its Chinese name, its Latin name and its medicinal part, or in the case of minerals, its class, family, ore name and main composition).

(b) Sample source: Authentic record of where the sample is from, such as whether it is collected from traditional places of origin or from places of origin with rich resources, or supplied by a GAP base; whether it is purchased from the place of origin, from the market or from agents, etc. This facilitates the purchase of raw materials for manufacturing, and ensures the traceability of test data.

(c) Time of collection and the person-in-charge: Sample collection time and the name of the person-in-charge.

(d) Investigation of goods supply: Information on whether the goods supply is adequate and stable, etc.

(e) Origin identification and the person-in-charge: Identify the variant based on the place of origin and the shape and form of the plants. For a plant of unknown origin, experienced specialists are to identify it from its appearance or microscopic characteristics. For closely related variants or wild variants that are hard to distinguish, one should study the fingerprints to make careful comparisons. If the fingerprints obtained have a high degree of similarity, these variants can be used but the details must be clearly recorded, and when practising GAP in the future, only one type of variant should be specified. If the fingerprints obtained have a poor degree of similarity, then one should identify the variants and use them for cultivation instead. For variant sources with poorly identified commercial products, one should narrow down the scope and gather resources so that only one type of variant is chosen and recorded in the pharmacopeia. In medicine formulation, the "emperor medicine" (君药) and medicinal materials of larger prescription quantities should be the focal point for variant identification, so as to avoid future difficulties that are unanticipated during the fingerprinting process.

(f) Quality assessment: To facilitate accurate analysis of the chromatograms, and to minimize judgment errors in the test results, the medicinal materials should meet pharmacopoeia or Ministry of Health regulations, and have accompanying detailed records. All medicinal material samples should be serial numbered, and photographs of the medicinal materials should be attached when necessary.

(g) Reserve samples, sample storage and labeling: As the study of HPLC fingerprints is a relatively lengthy process, the samples must be stored in a low-humidity and low-temperature area, away from direct sunlight. Labels should bear serial numbers. The serial number of the collected sample should be identical with the serial number of the storage sample and test sample. The number of reserve samples should not be fewer than three times that of the actual number of test samples, to ensure traceability if there are any discrepancies in the test results.

Preparation of Test Samples

In the preparation of test samples, every step of the process should follow standardized operations. For all the batches, the test sample preparation process should maintain consistency in order to ensure that the sample analysis is precise, accurate and enables reproducibility and comparability of the samples. The main operating procedure and data should be recorded in detail.

Sample Collection

Refer to the 2000 edition of the *Pharmacopoeia of the People's Republic of China* for the official Chinese method for Chinese medicinal material sample collection to ensure that the samples collected are representative and uniform. The laboratory samples collected should be consistent with the actual medicinal materials used. For above-ground samples, collect about 0.5–1 kg samples. Weigh the shoots, leaves, flowers and fruit separately and record the approximate ratio. For fruit-based medicinal materials, if the seeds are removed during actual use, the seeds should also be removed from the test samples, and it should be recorded as such. If the apparent mass of the medicinal material is not uniform (such as in the case of dissimilar size or dimension), one should note the representativeness of the sample collected. During subsequent testing, where necessary, one should carry out comparative testing to determine whether there are any significant differences in their compositions, to be used subsequently as a reference for practical applications.

Weighing of Sample

To follow conventional requirements, the selected samples must be suitably pulverized, then mixed well. Next, the required amount for testing must be weighed. The usual ratio of weighed sample to selected sample should be 1:10, that is, if 1 g of the test sample is required, it should be weighed out from 10 g of a uniformly mixed selected sample. Since a fingerprint needs to provide quantitative information, the usual requirement for weighing precision is to obtain three significant figures. The method of pulverization, the degree of pulverization and the amount of sample weighed depends on the nature of the medicinal material, the test sample preparation method and other actual conditions.

Sample Preparation

This is a crucial step in the sample preparation process. Depending on the physical and chemical nature of the chemical composition of the medicinal material, and the requirements of the test method, one should select an optimized extraction and separation method to prepare the test samples. During sample preparation, one should carry out a study on the different extraction solvents, extraction methods, purification methods, etc, making every effort to retain the chemical component of the test sample at the maximum limit. This is to ensure that the main chemical component of the Chinese medicinal material shows up in the fingerprints and to reflect the chemical component of the medicinal material in the fingerprints as much as possible. For Chinese medicinal materials from which only one or several components are extracted, extract and separate each chemical component according to their nature and prepare the fingerprints separately.

Preparation of Solution

The test sample solution should be prepared by dissolving the sample in a suitable solvent in an apparatus of specified volume, according to specified concentrations (g/mL or mg/mL).

Storage

As a normal requirement, the test solution should be freshly prepared as far as possible. If continuous testing is required, the test solution should be stored away from light, and at low temperature, in an airtight container for a short period, usually not exceeding two weeks. For solutions that are unstable, the storage period should not exceed 48 h.

Labeling

Clearly indicate the serial number or batch number on the test sample. It should be the same as the serial number of the collected sample or have a clear correlation to ensure that the data is traceable.

Preparation of Reference

To formulate a fingerprint, one should setup a reference material or reference peak, and depending on the nature of the component in the test sample, select a suitable control to be the reference. If there is no suitable control, one may select a suitable internal standard to be the reference. The reference should be prepared using a suitable method according to the requirements of the test method.

Choice of Reference

The reference for the fingerprint is usually selected from more than one main active components or target components, mainly for the purpose of checking the robustness and reproducibility of the fingerprints, and to help in the chromatographic identification. In a situation where there is no clear correlation between the drug and the clinical drug efficacy, the reference helps in identifying and evaluating the characteristics of the HPLC fingerprint. It is not equivalent to the control in quantitative analysis. Fingerprints are usually more complex, hence it is not easy to choose or spike in an internal standard. Fingerprints are not a form of quantitative analysis; hence, the purpose of the internal standard is also not equivalent to that of the internal standard in quantitative analysis. Therefore, it is important to carefully consider whether the use of an internal standard is necessary and feasible. If it is both necessary and feasible, one can consider spiking a suitable internal standard into the chromatogram. The reference should have its name, origin and purity listed. If there is no suitable reference available, one may select stable chromatographic peaks in the fingerprint as the reference peaks. Its chromatographic behavior and related data should be listed, and the chemical structure and the chemical name should be documented as far as possible.

Preparation of Reference

Weigh it accurately. Use a suitable solvent and make up the reference solution of specified concentration (g/mL or mg/mL).

Optimization of HPLC Conditions

The reason why HPLC has a broad range of applications is that it can be applied to medicinal materials that contain components such as alkaloids, glucosides, flavanones, organic acids, phenols and lignans. This is mainly because one is able to select the appropriate chromatographic conditions according to the test subject. The main HPLC conditions include the optimal choice of column, mobile phase and detector. One should setup the optimal HPLC conditions, so that the components of the medicinal material test sample can be resolved as much as possible. In other words, the more chromatographic peaks are obtained, the better it is. This will allow the internal characteristics of the Chinese medicinal materials to be fully manifested, providing sufficient information for medicinal material fingerprint evaluation and for its quality assessment.

Column

The column may be selected according to the main components of the Chinese medicinal materials. For Chinese medicinal materials that contain mainly alkaloidal compounds, the ion exchange column may be selected. For those that contain mainly polysaccharide compounds, a gel column may be selected. For those that contain mainly steroidal compounds, the C_{18} reversed-phase column may be selected. Currently, the column that is most commonly used in medicinal material fingerprint study is the C_{18} bonded-phase type of reversed-phase ODS column. The normal-phase column, which is rarely used, is for the separation of homologous or isomeric compounds. The ion exchange column is used for the separation of water-soluble ionic compounds. Some compounds may require the use of an amino column.

Mobile Phase

The choice of mobile phase should depend on the actual conditions, and the solvent system with the most optimized method should be used. The best separation conditions should satisfy the following requirements:

1. All the components of the test sample can be analyzed or the sample can be analyzed into as many components (peaks) as possible.

2. All the components in the test sample can be separated satisfactorily.
3. The analysis time is as short as possible.

Optimization of the separation conditions mainly involves optimization of chemical factors such as mobile phase composition, pH value of mobile phase and ion pair reagent concentration. There are many optimization methods; the most representative and relatively more developed methods are the triangle methods (including the Glajch triangle method, tetrahedron method and prism method), graphical methods (including the window diagrams method and overlapping resolution mapping) and direct methods (including the simplex method, complex method and repeated design method). The general consensus is that the triangle methods and the direct methods are suitable for the analysis of both known and unknown samples, while the graphical methods are only suitable for the analysis of known samples.

For the mobile phase, other than optimization of the solvent composition, dynamic changes in the solvent ratio should also be considered. As Chinese drugs contain many complex components, normal isocratic elution is unable to achieve the separation of components of different natures, giving rise to fewer peaks in the chromatogram and insufficient information required for fingerprint evaluation. Under most circumstances, gradient elution should be used. Under suitable gradient conditions, compounds of widely differing natures are all separated. The more chromatographic peaks are obtained, the better it is for the fingerprint evaluation and quality assessment of the medicinal material.

Detector

Currently, the detectors available for selection include ultraviolet–visible light (UV-Vis) detector, diode array detector (DAD), fluorescence detector (FD), electrochemical detector (ECD), refractive index detector (RID), evaporative light-scattering detector (ELSD).

Of these, the UV-Vis detector is the most commonly used detector for HPLC. Its degree of sensitivity, precision and linearity range are relatively good. It is suitable for component detection of compounds with π-π conjugated or p-π conjugated structure, while a large number of compounds that are without double bonds are not detected. The DAD is a type of multiple-channel optical system and a new model of UV detector with high resolution. It can simultaneously derive the sample chromatogram and the absorption spectra of every chromatographic component (A-λ curve), to obtain the 3-D spectrum fingerprint chromatograms, and is more suitable for test objects of a relatively complex composition that absorb UV light. This detector is able to evaluate the purity of the HPLC peak, and hence improve the scientific accuracy of the test results. ELSD is able to detect substances that do not absorb UV light, and is suitable for the detection of steroidal saponins and polysaccharide compounds present in Chinese drugs. For those compounds that only exhibit end (lower-wavelength) absorption, such as ginsenoside and astragaloside IV, ELSD is a more suitable detector.

Sample Analysis and Method Validation

Chinese medicinal material samples are prepared using optimized extraction and separation methods, and the analysis should be carried out under the best chromatographic conditions for separation and analysis. For those Chinese medicinal materials that contain the same type of or similar components, one may prepare a single fingerprint. For those that contain complex components, a single fingerprint is not able to completely reflect the intrinsic property of that medicinal material. Through extensive comparison and testing, fingerprints that can sufficiently represent the characteristics of Chinese medicinal materials are obtained, to satisfy the fingerprint requirements of specificity, reproducibility, stability and universal applicability. To confirm the reliability of the test results, the stability of the test sample, the precision of the instrument and the reproducibility of the test method must go through rigorous method validation.

Stability Test

This is done mainly to validate the stability of the test sample. Using the same test sample, the sample is tested at different time intervals (0 h, 1 h, 2 h, 4 h, 8 h, 12 h, 24 h, 36 h and 48 h). The uniformity of the relative retention times and the peak area ratios of the chromatographic peaks are studied and the time of analysis determined.

Precision Test

This is to validate the precision of the instrument. The same sample is injected at least five times consecutively. The uniformity of the relative retention times and the peak area ratios are studied. In the fingerprint, for each peak that defines the peak area ratio of the common peaks, the relative standard deviation (RSD) of its peak area ratio should not be greater than 3%. The relative retention time of each peak should be within ±1 min of the average retention time.

Reproducibility Test

This is to validate the reproducibility of the test method. At least five samples from the same batch number are used. The test samples are prepared according to the selected extraction and separation methods. The samples are tested under selected HPLC conditions. The uniformity of the relative retention times and peak area ratios of the peaks are studied. In the fingerprint, for each peak that defines the peak area ratio of the common peaks, the RSD of its peak area ratio should not be greater than 3%. The relative retention time of each peak should be within ±1 min of the average retention time.

HPLC Fingerprint Construction

Based on the number of peaks, peak area (integrated value), peak position (retention time) and other related parameters obtained from the test results of an adequate sample number (more than 10 batches), the common peaks (relative retention time, peak area ratio) are determined. The characteristic peaks (peak combination) are then selected and the fingerprint constructed. Arabic numerals are used to label the common peaks, and "S" is used to label the reference peak. During the experiment, the 2 h chromatogram should be recorded so as to observe the profile of the chromatographic peaks after 1 h.

The fingerprints of Chinese medicinal materials should be fully representative and specific. Analysis and comparison of representative samples with various places of origin, different grades and specifications or different seasons of collection, etc, should be carried out. From there, the common peak with a relatively constant peak area is identified as the characteristic peak for the Chinese medicinal material. The selected characteristic peaks should possess specificity. For Chinese medicinal materials from more than one source, the specific differences among the variants must be studied.

Analysis and Evaluation of Fingerprints

Critical parameters (common peaks, overlapping ratio, *N* strong peaks, characteristic fingerprint, etc) for analyzing and comparing fingerprints are established according to the information obtained from the fingerprints. The degrees of similarity and difference among the characteristic fingerprints are calculated for the fingerprint evaluation. To develop a feasible standard for quantitative evaluation using HPLC fingerprinting, computer technologies (principal component analysis, clustering analysis, the artificial neural network method or similarity degree analysis, etc) are utilized to analyze fingerprint information and to evaluate the degree of similarity among chromatograms.

Construction of Important Parameters

Common peaks

Common peaks are identified as those peaks with the same relative retention time in the HPLC chromatograms of different test samples.

Standardization of common peaks

The retention time of the reference is used to calculate the relative retention times of the fingerprint peaks. Common peaks of the Chinese medicinal materials are standardized according to the results of over 10 batches of samples.

Ratios of common peak areas

If the fingerprint of the control is used as the reference fingerprint, the ratios of the common peak areas to the reference peak area are calculated by designating the peak area of the reference as 1. If an internal standard is used as the reference, then one of the common peaks (which should be relatively larger and more stable) is used as the reference peak, and the ratios of the common peak areas to the reference peak area are calculated by designating the peak area of the reference as 1. The ratios of the common peak areas should be relatively stable. The ratio of common peak areas in the test samples of the Chinese medicinal material is compared with the ratio of common peak areas in the standard fingerprint. If a single peak area is more than or equal to 20% of the total peak area, the peak area of that peak in the test sample should not deviate more than ±20% from that in the standard fingerprint. If a single peak area is more than or equal to 10% or less than 20% of the total peak area, the difference should not be more than ±25%. If a single peak area is less than 10% of the total peak area, there is no requirement on its peak area ratio. However, the relative retention times of all common peaks should be standardized. If the common peaks do not achieve baseline separation, the total peak area of this group of peaks should be calculated and used as the peak area. At the same time, the relative retention time of each peak in this group should be standardized.

Areas of non-common peaks

After comparing the chromatograms of the test samples of the Chinese medicinal materials with the standard fingerprint, the peaks with different relative retention values (those peaks which are not common peaks) are classified as non-common peaks. The total area of non-common peaks should not be more than 10% of the total peak area.

Overlapping ratio

The overlapping ratio is twice the total number of common peaks divided by the sum of the number of peaks in the HPLC fingerprint of a Chinese medicinal material and in the chromatogram of the test sample, expressed as a percentage. The formula is

$$\text{Overlapping ratio} = \frac{\text{Number of common peaks} \times 2}{\text{Number of peaks in the fingerprint} + \text{Number of peaks in test sample}} \times 100\%$$

The overlapping ratio indicates the similarity among the fingerprint chromatograms. The higher the overlapping ratio, the greater the similarity among the fingerprints. Depending on the scenario, a reasonable working range for the overlapping ratio should be specified. The overlapping ratio is an important qualitative parameter which provides reliable evidence for identification of the variants of Chinese medicinal materials.

N strong peaks

N strong peaks should be identified based on the actual peaks presented and peak area values that emerged. First, the sizes of the peak areas are listed in descending order and the first *N* peaks are selected as strong peaks. The total peak area of the strong peaks should be more than 70% of the total peak area. The value of *N* depends on two factors. The first is the total number of peaks in the chromatogram. The value of *N* should fall between 1/5 and 1/3 of the total peak numbers. The second is the total peak area of the *N* strong peaks. In addition, the frequency and the position of the different *N* strong peaks in the test sample of the Chinese medicinal material should be considered. *N* strong peaks indicates the relative amount of the main components in the Chinese medicinal materials, which provides important information and a basis to assess the quality of Chinese medicinal materials.

Characteristic fingerprint

Common peaks are chromatographic peaks that exist in the chromatograms of all the test samples of a Chinese medicinal material, and can be considered a unique characteristic of the materials. That is why they are also called characteristic fingerprint peaks. A characteristic fingerprint is made up of a fixed group of characteristic fingerprint peaks that reflects the multi-component characteristic of a Chinese medicinal material. It provides more detailed information and evidence for the identification of a variety of Chinese medicinal materials, especially for the differentiation of medicinal materials of different species within the same genus or different species that contain the same active ingredients.

The establishment of the characteristic fingerprint requires all the protocols to be standardized. The test samples should be prepared with optimized extracting and separating methods. The tests should be conducted with the best chromatographic conditions for separation and analysis. The characteristic fingerprint peaks are standardized using related parameters from the results of over 10 batches of testing. A series of characteristic fingerprint peaks is selected to construct the characteristic fingerprint of a Chinese medicinal material.

Degrees of similarity and difference among characteristic fingerprints

Degrees of similarity and difference are used to quantify the similarity between the chromatograms of the test sample and the standard fingerprint. They combine qualitative information with quantitative data, which can be used as an important basis for final identification.

The degree of similarity between the chromatograms of the test sample and the standard fingerprint is calculated based on the characteristic fingerprint pattern. It indicates the similarity in the fingerprint patterns of these two chromatograms. The first step is to calculate the degree of similarity of every characteristic fingerprint peak. The formula is

$$\text{Degree of similarity of a characteristic fingerprint peak} = \frac{A_I^T}{A_I^F} \times 100\%$$

Then the total degree of similarity of the characteristic fingerprint is obtained by calculating the summation of the degrees of similarity of all the characteristic fingerprint peaks. The formula is

$$\text{Total degree of similarity of the characteristic fingerprint} = \frac{\Sigma\left(\dfrac{A_I^T}{A_I^F}\right)}{n}$$

In those formulae,

A_I^T is a characteristic fingerprint peak area in the chromatogram of the test sample,

A_I^F is a characteristic fingerprint peak area in the standard fingerprint,

n is the total number of the characteristic fingerprint peaks, and

$\sum$ is the summation symbol.

The total degree of similarity of the characteristic fingerprint should be equal to or less than 1. If the degree of similarity is 1, both of them are completely the same. If the degree of similarity is less than 1, the larger the value, the higher the degree of similarity between them. Any value more than 1 is invalid.

The degree of difference is the difference in peak area between a characteristic fingerprint peak in the test sample and its corresponding characteristic fingerprint peak in the fingerprint chromatogram, divided by the peak area value of the characteristic fingerprint peak in the fingerprint chromatogram.

The total degree of difference of the characteristic fingerprint is obtained by calculating the summation of the degrees of difference of all the characteristic fingerprint peaks. It indicates the degree of difference between these two chromatograms of the characteristic fingerprints. The formulae are as follows:

$$\text{Degree of difference of a characteristic fingerprint peak} = \frac{\left|A_I^F - A_I^T\right|}{A_I^F}$$

Total degree of difference of the characteristic fingerprint =

$$\frac{\Sigma\left(\dfrac{\left|A_I^F - A_I^T\right|}{A_I^F}\right)}{n}$$

Computer Analysis of Fingerprints

Fingerprint information

After Chinese medicine samples have gone through sample preparation and chromatographic separation and analysis, the fingerprint data that characterize the chemical components in Chinese medicinal materials are obtained. In the process, data related to characteristic features of a Chinese medicinal material should be collected. The parameters chosen for this purpose can be relative retention time and relative peak area.

For HPLC fingerprints, especially the high-resolution ones, the chromatograms should be prepared from representative samples. Significant characteristic peaks in stable fingerprints should be selected and their peaks should be numbered. Main characteristic peaks should separate well from their neighboring peaks with an effective resolution of over 1.2. Other characteristic peaks should be separated to a certain degree. The peak-to-valley distance should be at least 2/3 of the peak height. If the distance is less than 2/3 of the peak height, the two peaks will be calculated as one peak. To allow the identification and matching of characteristic peaks, the variation of the relative retention time of a peak should be controlled to be between ±5% and ±10%. To study the stability and repeatability of a fingerprint, either a reference should be used or an appropriate peak in the same fingerprint should be selected as the internal standard for calculating the relative retention time and relative peak area ratio (A_i/A_s) of characteristic peaks. The relative retention time and relative peak area ratio are used as parameters. If a peak area exceeds 10% of the total peak area, its relative peak area value will be specified. This will be used to establish the reference fingerprint, which will be used to evaluate the authenticity of a sample.

Analysis of fingerprint information

HPLC fingerprints of Chinese medicinal materials are very complicated. They contain information which may or may not be useful for classification. Manual analysis may not be accurate. If the method is supplemented with computer analysis and identification techniques, its applications will be expanded significantly. Fingerprint analysis means analyzing the fingerprint, selecting useful characteristic information (i.e. characteristic information which can reflect differences in the chemical components and their concentrations in different samples), and converting it into a digital matrix, which can be recognized by the computer, to obtain the objective criteria for evaluating the quality of Chinese medicinal materials and to provide standards for establishing a Chinese medicine quality evaluation system. Currently, there is no universal standard for fingerprint analysis. An appropriate method for chemical pattern identification should be selected according to the situation. Below are the most advanced and current computer-assisted fingerprint analysis methods.

Fuzzy information analysis

This method is used for classification, identification or deduction from fuzzy information. The fuzziness of the information refers to the uncertainty in the classification process, which is caused by the intermediary transition of objective differences. The intermediary transition of a difference means the process of quantitative change to qualitative change from one side of the change (such as genuine herb) to the other side of the change (such as counterfeit herb). Without a doubt, there is a certain degree of fuzziness in the characteristics of the fingerprints of Chinese medicines. This kind of fuzziness is usually represented by a membership grade, which is used for Chinese medicine authentication. Usually, the characteristics of fuzzy information can be discovered by comparing the definite information and random information. Inexorable laws can be obtained by analyzing definite information, while statistical laws can be obtained by analyzing random information. The characteristic of fuzzy information is that it can provide a fuzzy reference for people to fuzzily identify its corresponding inexorable laws and statistical laws. For authentication of Chinese medicinal materials, the basic procedures for computer-assisted fuzzy identifications are

1. Characteristic information collection: It means collecting digitized characteristic information from the fingerprints of a Chinese medicinal material, which represents its quality. When collecting the characteristic information, the following rules should be followed: Discard, as much as possible, information that will not affect the determination of classification, and keep all the information that will affect the determination of classification. Otherwise, the accuracy of classification will be affected.
2. Follow a certain method to determine the subordinate degree of the characteristics of a Chinese medicine fingerprint for genuine modules and counterfeit modules.
3. Authenticate the Chinese medicinal materials by following the maximum subordinate principle or threshold principle.

Artificial neural network method

This is a brand new information-processing method, which mimics the functions of human brains. Its theory is based on the mathematical model of a neural network. A neural network system contains a lot of basic information-processing units, which are called neurons or ganglia.

There are plenty of complex connections among neurons. The strength of a connection can be represented by a variable weightage. Neurons have learning capacities. Based on certain learning rules, they can change the strength of connections according to problems being processed. A neuron network has its special ability in dealing with problems involving indeterminate relationships, backgrounds and rules. It is extremely suitable for processing the fingerprint information of Chinese medicinal materials.

The back propagation (BP) model is currently the most widely used artificial neural network model. It has several advantages, such as high reliability and high speed of classification. A BP network has multiple connection nodes. It consists of an input layer, a hidden layer and an output layer, which correspond to the input, transition and output of the system, respectively. There are two steps in the BP network's operation process. First, a training set is formed from a group of samples with known results. For every input of a sample, an output result is obtained according to forward propagation. Second, the difference between the actual output result and expected output result is calculated. The difference will be fed back from the output layer to the input layer to modify the connection weightages of different connection nodes. This procedure is repeated again and again until the differences among every model and every output unit in the whole training set are less than a certain value. For a trained network, the system rule, prediction ability and variable transformation are incorporated in the network in the form of weightages. Processed results will be obtained when information is imported into an input layer of the trained network.

Grey relational grade cluster method

The grey system is a system that contains both known and unknown information. As the fingerprints of Chinese medicinal materials have grey characteristics, they can be processed by the grey relational grade method of the grey system theory. In this method, a new pattern recognition model is established by measuring defined relative correlation. Supposing that there are several Chinese medicinal materials, since every sample has several quantitative indices of characteristics of fingerprints, the evaluation unit sequence is composed. To use the grey relational grade as the evaluation measurement, the reference sequence should first be selected. Then the relational grade (r_s) obtained by comparing it with the optimal reference sequence and the relational grade (r_t) obtained by comparing it with the worst reference sequence are calculated. The higher the r_s, the stronger the correlation between the evaluation unit sequence and the optimal reference sequence, indicating a better evaluation unit. The reverse is true for the r_t value. The lower the r_t, the better the evaluation unit. An ideal evaluation unit should have the highest correlation with the optimal reference sequence and at the same time have the lowest correlation with the worst reference sequence. Thus, the relative relational grade (r_i) compared with both the optimal reference sequence and the worst reference sequence is defined as

$$r_i = \frac{r_s}{r_s + r_t}$$

Obviously, the higher the r_i, the better the evaluation unit. From the value of the relative relational grades of different evaluation units, each evaluation unit can be placed in sequence based on merit. Then the qualities of different Chinese medicinal materials can finally be determined.

Principal component analysis

The principle of the analysis is to factorize the singular values of a fingerprint data matrix into two orthogonal matrices (**U**, **V**) and a diagonal matrix (**S**). The product of the two orthogonal matrices (**U**, **V**) multiplied by the diagonal matrix (**S**) is then calculated. Two of the three values are used as the ordinate and abscissa in the projection diagram. The other value is the projection point in the projection diagram. The changing patterns of the principal components can be identified according to the different places the projection points are located. Samples with a lower total quantity are located to the left end of the main component projection diagram. As the position of the sample point is shifted from the left to the right part of the principal component projection diagram, the total peak area in the whole fingerprint becomes bigger and bigger. There is a frequently encountered problem during fingerprint analysis, that is, how to reflect the total amount of extracts of a fingerprint during the evaluation of degrees of similarity. Obviously, it is difficult to reflect the total amount, whether using the correlation coefficient or vector angle cosine. This is due to the fact that when measuring a similarity degree, only the overall similarity of a fingerprint is compared and the differences in the extract quantities are not considered. For example, the chemical pattern recognition method (main component analysis) used in the Evaluation Software for Chinese Medicine (Computer-assisted Similarity Evaluation System) is to project those values into a lower-dimensional space to analyze their nuances, which makes a proper evaluation possible.

Evaluation of a fingerprint's degree of similarity

A fingerprint represents all the characteristics of a sample, which can be used to describe the relationships among samples, visually compare the peak number, peak sequences, peak areas (or peak heights) and relative proportions of the peaks to see if they are similar. All these are just subjective descriptions. The degree of similarity is obtained by measuring the similarity between the fingerprint of a Chinese medicine test sample and the fingerprint of a reference. Comparing the degree of similarity is a kind of fuzzy information analysis method which can quantify the similarity among fingerprints using digitized descriptions that are more objective.

Methods and principles of calculating degree of similarity

Euclidean distance measurement, included angle cosine measurement, correlation coefficient measurement, similarity index measurement and similarity ratio measurement, are often used for calculating the degrees of similarity of Chinese medicine test samples. Measurements of the correlation coefficient, included angle cosine, similarity index and similarity ratio are used to determine the similarities in the changing patterns of characteristic variables. They are good for providing information on the closeness in relationships among test samples during the authentication of Chinese medicinal materials. The Euclidean distance reflects the closeness in relationships among the test samples that have differences in the characteristic variables. This method is good for the quality control of Chinese medicinal materials.

1. Euclidean distance: Similarity reflects the relationship among investigated subjects and can be measured by distance. The frequently used method is Euclidean distance, or the so-called second-order Minkowski measure.

$$d_{ir} = \sqrt{\sum_{k=1}^{m} (X_{ik} - X_{rk})^2}$$

   In this formula, X_{ik} represents the kth characteristic variable (k = 1, 2, ..., m) of the ith sample, X_{rk} represents the kth characteristic variable (k = 1, 2, ..., m) of the mean vector of a common pattern.

   During the calculation of the Euclidean distance, the absolute value of the absolute distance is replaced by a square function, which makes the calculation easier and can emphasize the effects from bigger characteristic variables. The Euclidean distance places extra emphasis on the differences among values of characteristic variables. It does not consider the changing patterns of characteristic variables, i.e. the similarities of those changing patterns.

2. Correlation coefficient: Pearson's correlation coefficient (i.e. simple correlation coefficient) is used to measure the degree of similarity among test samples in cluster analysis. In fingerprints, the degree of similarity measured by the correlation coefficient is

$$r_{ir}=\frac{\sum_{k=1}^{m}(X_{ik}-\overline{X}_i)(X_{rk}-\overline{X}_r)}{\sqrt{\sum_{k=1}^{m}(X_{ik}-\overline{X}_i)^2\sum_{k=1}^{m}(X_{rk}-\overline{X}_r)^2}}$$

where X_{ik} represents the kth characteristic variable of the ith sample, $\overline{X}_i$ represents the mean value of all variables of test samples, X_{rk} represents the kth characteristic variable of a common pattern, and $\overline{X}_r$ represents the mean value of all variables of a common pattern.

The correlation coefficient is not related to variable units and is not sensitive to the values of different characteristic variables. The different values of different variables, which have little effect on the changing patterns of characteristic variables, are ignored. The correlation coefficient, also called shape measurement, is to measure the similarities among similar shapes in the changing patterns of characteristic variables of different samples. The degrees of similarity obtained in this way can be used for authenticating Chinese medicinal materials and providing qualitative information.

3. Included angle cosine: Inspired by the similar figures in geometry, the degree of similarity of the included angle cosine between the characteristic vector of a test sample's fingerprint and vector of the common pattern is calculated based on the multi-dimensional vector included angle. This kind of degree of similarity is related to the changing patterns of characteristic variables in fingerprints, which can provide information for the authentication of Chinese medicinal materials.

$$C_{ir}=\frac{\sum_{k=1}^{m}X_{ik}\bullet X_{rk}}{\sqrt{(\sum_{k=1}^{m}X_{ik}^2)(\sum_{k=1}^{m}X_{rk}^2)}}$$

4. Similarity index: Degrees of similarity can be measured by the similarity index when both the mean vector and the standard

deviation vector of a common pattern are considered. This kind of degree of similarity is related to the changing patterns of characteristic variables in fingerprints.

$$C_{ir} = \frac{1}{m}\sum_{k=1}^{m} e^{-\frac{3}{4}\frac{(X_{ik}-X_{rk})^2}{S_{rk}^2}}$$

5. Similarity ratio: Degrees of similarity measured by the similarity ratio are related to the changing patterns of characteristic variables in fingerprints. The mathematical concept of the similarity ratio is simple and its processing procedures are simple, convenient and fast.

$$C_{ir} = \frac{\sum_{k=1}^{m} X_{ik} \bullet X_{rk}}{\sum_{k=1}^{m} X_{ik}^2 + \sum_{k=1}^{m} X_{rk}^2 - \sum_{k=1}^{m} X_{ik} \bullet \sum_{k=1}^{m} X_{rk}}$$

Software for evaluating degree of similarity

In current practical work, most degrees of similarity are calculated by computer-aided similarity evaluation software for Chinese medicine fingerprints, which is recommended by the Chinese Pharmacopoeia Commission. There are two types of evaluation software; one is The Fingerprint Analysis System of Chinese Medicine (Professor Cheng Yiyu of Zhejiang University), and the other is Computer-assisted Similarity Evaluation System (Professor Liang Yizeng of Central South University). Both use fuzzy information analysis. Degrees of similarity are calculated using the included angle cosine method. Every fingerprint can be regarded as a group of peak area values or data points at their corresponding retention time. If this group of values are regarded as vectors in a multi-dimensional space, the similarity between two fingerprints is now converted into the similarity between two vectors in a multi-dimensional space. *Cos* θ is used for quantitatively determining and representing the similarity between fingerprints. The closer *cos* θ is to 1, the more similar the two vectors are.

If there are N peaks in a fingerprint, it can be expressed in an

N-dimensional vector space. If the reference fingerprint is expressed as $X_0 = [X_{01}, X_{02}, \ldots, X_{0N}]$ and X_{0i} is the peak area value of peak i, then the fingerprint for testing is expressed as $X = [X_1, X_2, \ldots, X_N]$. The fingerprint of the reference and the fingerprint of the test sample are expressed as two points in the N-dimensional vector space. The overall degree of similarity of these two fingerprints is evaluated by calculating their similarity from the included angle cosine of these two points. For example,

$$S(x_0, x) = \frac{\sum_{i=1}^{N} X_{0i} \bullet X_i}{\sqrt{\sum_{i=1}^{N} X_{0i}^2}\sqrt{\sum_{i=1}^{N} X_i^2}}$$

Except for a few species, which should be decided on a case-by-case basis, most of the degrees of similarity of Chinese medicine fingerprints are considered to have met the criteria when the calculated results are within the range of 0.9–1.0 (or expressed as 90–100). If the similarity degree of a Chinese medicinal material is less than 0.9, and when it is difficult to reject the samples just by visual comparison, further evaluation by chemical pattern recognition methods (such as principal component analysis) may be performed.

The detailed protocol for computer-aided similarity evaluation software for Chinese medicine fingerprints is as follows: First, data of statistically significant samples (.txt, .csv file) is imported for establishing common patterns. The data is pre-processed and the resized fingerprints are compared. Peak positions (retention times) are adjusted and peaks are automatically matched. The degrees of similarity of the fingerprints are calculated. After repeatedly deleting and adding samples, the figure and data of a common pattern are obtained and the common pattern template is created. Then test samples (.txt, .csv file) required in the similarity calculation software are generated at the workstation and are processed by the similarity calculation software.

Protocol for Similarity Degree Calculation Software for Traditional Chinese Medicine Fingerprints: Data is imported (.txt file). The data is pre-processed and the resized fingerprints are compared. Peak positions (retention times) are adjusted and peaks are automatically matched. The degrees of similarity of the fingerprints are calculated. Results are exported and reports are printed.

Protocol for Computer-assisted Similarity Evaluation System: Data is imported. The data is pre-processed, which includes data compression and translation. Peaks are identified and matched. The degrees of similarity are calculated. Principal component analysis can be used to further evaluate different amounts of a certain component.

References

1. 任德权. (2001) 中药指纹图谱质控技术的意义与作用. 中药新药与临床药理, **12**(3):135–140.
2. 谢培山. (2001) 中药色谱指纹图谱鉴别的概念、属性、技术与应用.中国中药杂志, **26**(10):653–655.
3. 梁逸曾. (2001) 浅议中药色谱指纹图谱的意义、作用及可操作性. 中药新药与临床药理, **12**(3):196–200.
4. 谢培山. (2001) 中药色谱指纹图谱质量控制模式的研究与应用—若干实质性问题的探讨 (一).世界科学技术—中药现代化, **3**(3):18–23.
5. 郑颖, 吴凤锷. (2003) 中药指纹图谱的研究进展.天然产物研究与开发, **15**(1):55–60.
6. 王永刚, 吴忠, 魏凤环, 等. (2003) 中药指纹图谱研究的现状与未来. 中药材, **26**(11):820–825.
7. 刘朝槃, 王冬梅, 白洁, 等. (2003) 色谱技术在中药指纹图谱研究中的应用. 色谱, **21**(6):572–576.
8. 王天松. (2000) 毛细管电泳与中草药指纹图谱. 中成药, **22**(6):397–399.
9. 沈平孃. (2001) 高速逆流色谱 (HSCCC) 技术与色谱指纹谱. 中成药, **23**(5):313–315.
10. 罗国安, 王义明, 曹进. (2000) 多维多息特征谱及其应用. 中成药, **22**(6):395–397.
11. 苏薇薇, 吴忠, 全健. (2001) 中药指纹图谱的构建及计算机解析. 中药材, **24**(4):295–298.
12. 赵宇新, 李曼玲. (2002) 模式识别在中药质量评价中的应用进展.中国中药杂志, **27**(110):808–811.
13. 王玺, 王文宇, 张克荣, 等. (2003) 中药HPLC指纹图谱相似性研究的探讨. 沈阳药科大学学报, **20**(5):360–362.
14. 国家药品监督管理局. (2000) 中药注射剂指纹图谱研究的技术要求(暂行).
15. 蔡宝昌, 潘扬, 殷武. (2000) 指纹图谱在中药研究中的应用. 世界科学技术—中药现代化, **2**(5):9–14.

HPLC Fingerprints of Different Herbs

2.1

Renshen

人参

Radix et Rhizoma Ginseng

Introduction

Radix et Rhizoma Ginseng is the dried root and rhizome of *Panax ginseng* C.A. Mey. (family Araliaceae). It is mainly produced in Jilin, Liaoning and Heilongjiang Provinces. There are two types: one is cultivated Yuanshen (园参), the other is wild Shanshen (山参). Commercially available products are mainly Yuanshen (园参).

According to studies on resources, *Radix et Rhizoma Ginseng* is planted abundantly in Fusong, Ji'an, Jingyu, Dunhua, Antu, Changbai, Tonghua, Linjiang, Shulan, Helong, Yanji, Huichun, Wangqing, Jiaohe, Huadian, Hunjiang, Hunnan, Liuhe, Panshi, Yongji and Hailong of Jilin Province. Among them, Fusong, Ji'an, Jingyu, Dunhua and Antu are the main producing areas of *Radix et Rhizoma Ginseng*.

Chemical Composition

Radix et Rhizoma Ginseng mainly contains saponins, volatile components, fat-soluble components (panaxynol, panacene, sterol, fatty acid) and polysaccharides. Its main active ingredients are saponins, which mainly are ginsenosides Ro, Ra_1, Ra_2, Ra_3, Rb_1, Rb_2, Rb_3, Rc, Rd, Re, Rf, Rg_1, Rg_2, Rh_1, Rh_2, Rs_1 and Rs_2. There are three types of sapogenins, which are protopanaxadiol, protopanaxatriol and oleanolic acid. The chemical structures of ginsenoside are listed on page 78.

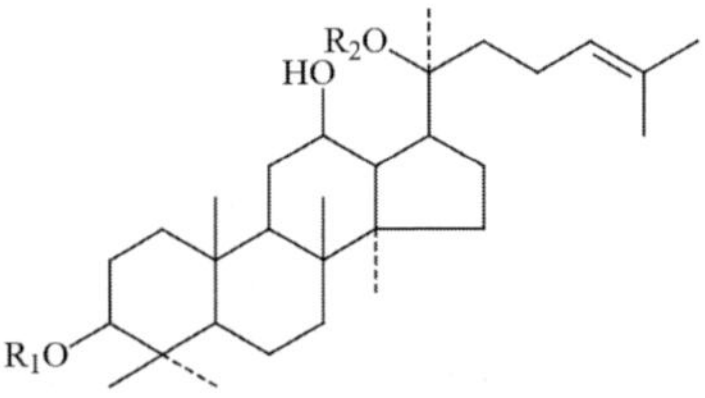

Protopanaxadiol $R_1=R_2=H$

Ginsenoside	R_1	R_2
Ra1	glu2–1glu*	glu6–1ara(P) 4–1xyl
Ra2	glu2–1glu	glu6–1ara(F) 2–1xyl
Ra3	glu2–1glu	glu6–1glu 3–1xyl
Rb1	glu2–1glu	glu6–1glu
Rb2	glu2–1glu	glu6–1ara(P)
Rb3	glu2–1glu	glu6–1xyl
Rc	glu2–1glu	glu6–1ara(F)
Rd	glu2–1glu	glu
Rg3	glu2–1glu	H
Rh2	glu	H
Rs1	glu2–1glu6–1ac	glu6–1ara(P)
Rs2	glu2–1glu6–1ac	glu6–1ara(F)

* glu is glucose, ara is arabinose, xyl is xylose, ac is acetyl, *sic passim.*

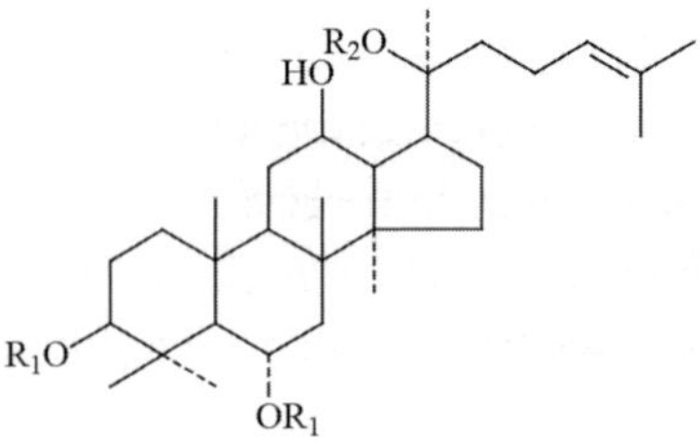

Ginsenoside	R_1	R_2
Re	glu2–1rha**	glu
Rf	glu2–1glu	H
20-glu-Rf	glu2–1glu	glu
Rg_1	glu	glu
Rg_2	glu2–1rha**	H
Rh_1	glu	H

Protopanaxatriol $R_1=R_2=H$

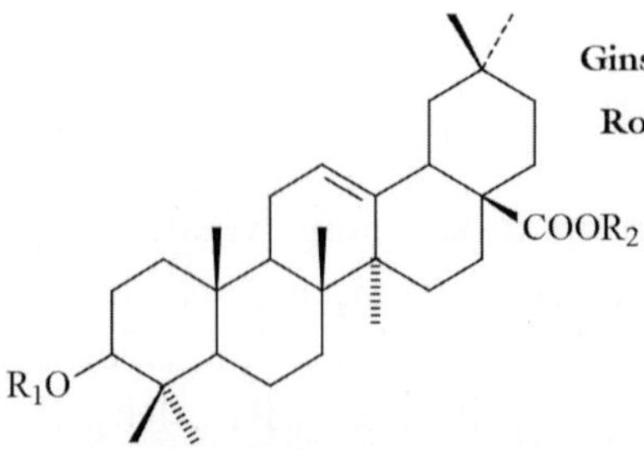

Ginsenoside	R_1	R_2
Ro	R_1=glucuronate2–1glu	R_2=glu

Oleanolic acid $R_1=R_2=H$

** rha is rhamnose, *sic passim.*

Experiment Materials

Samples

Radix et Rhizoma Ginseng samples were collected from Renshen Farm in Changbaixian Baoquanshan of Jilin Province in September 2000. The samples were five-year-old cultivated species and were processed to sun-dried ginseng specifications. There were 12 samples with variant identifications completed on all of them.

Reference

Ginsenoside Re used for quantitative analysis was provided by the National Institute for the Control of Pharmaceutical and Biological Products (China). The lot number is 0754-9608.

Reagents

Acetonitrile (chromatographic-grade, Tedia Company Inc., USA). Methanol (analytical-grade methanol was purchased from Shanghai Chemical Reagents Pte. Ltd.; chromatographic-grade methanol was purchased from Tedia Company Inc.). Double-distilled deionized water (prepared by our lab).

Preparation of Samples

2 g of crude powder of the sample was accurately weighed. After adding 100 ml of 80% ethanol, it was extracted twice by reflux for 3 h each time. The ethanol solutions were combined and ethanol was recovered. The residue was dissolved by 20 ml of water and was defatted three times by an equivalent amount of ethyl acetate. Ethyl acetate was then evaporated using a water bath and the pH was adjusted to 7.0 by adding ammonia solution. It was extracted by *n*-butyl alcohol (which was saturated with an equivalent amount of water) five times. The *n*-butyl alcohol solutions were combined and *n*-butyl alcohol was

recovered. The residue was dissolved in 20% acetonitrile and topped up to 10 ml. After being filtered, it was then ready to be used as a test sample.

Preparation of Reference

1 mg of ginsenoside was accurately weighed, dissolved and topped up to 1 ml by adding 20% acetonitrile in a volumetric flask. After shaking, it was mixed well and ready to be used as a reference.

HPLC Conditions

Instrument: Agilent 1100 high-performance liquid chromatography, including quaternary pump, online degasser, column temperature chamber, variable wave detector and chemical workstation (chemstation system workstation)
Column: ZORBAX SB-C_{18} (4.6 mm × 250 mm, 5 μm)
Mobile phase: Acetonitrile: water (gradient elution)
Flow rate: 1.0 ml/min
Column temperature: 30°C
Detecting wavelength: 203 nm
Run time: 70 min
Injection volume: 20 μl sample, 20 μl reference

Methods

The sample solution was injected into the liquid chromatograph. The chromatogram was obtained under selected HPLC conditions. The experimental method was validated.

Mobile Phase Selection

A comparison of experimental results showed that when acetonitrile: water was used as the mobile phase, gradient elution could successfully separate the peaks and produce the maximum number of peaks. The duration of the analysis was 70 min. The 2 h chromatogram of the samples showed no characteristic peaks after 70 min.

Method Validation

1. Stability: Samples from the same batch were tested at 0 h, 2 h, 8 h and 24 h. The relative retention time was stable. The RSD of the relative peak areas of the main common peaks was 1.45%.
2. Precision: Five consecutive injections of the samples (from the same batch) were conducted to determine the precision. The RSD of the relative peak areas of the main common peaks was 1.13%, which met the requirement.
3. Reproducibility: Five samples from the same batch were prepared and tested. The RSD of the relative peak areas of the main common peaks was 0.86%, which met the requirement.

Sample Analysis

The HPLC fingerprints of the samples from 10 batches were analyzed under the conditions described above.

Establishment of Fingerprints

The number of peaks, peak area and relative retention time of 10 batches of samples were analyzed and compared to obtain an optimized fingerprint (Fig. 1.1).

Results and Discussion

Common Peak Standardization

Common peaks were assigned according to the results of 10 batches of samples. There were 17 common peaks in the HPLC fingerprints

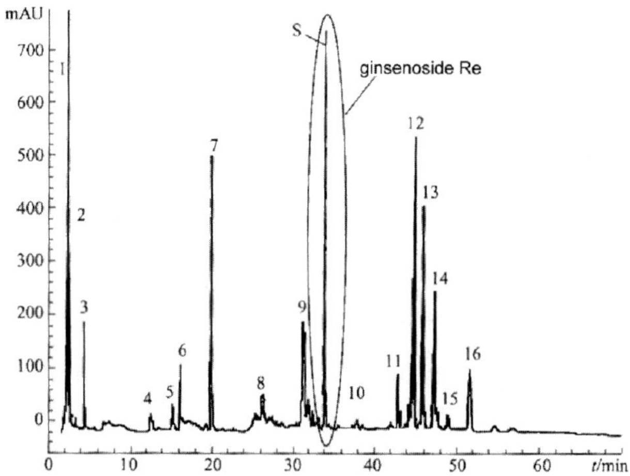

Fig. 1.1. HPLC fingerprint of Renshen *(Radix et Rhizoma Ginseng)*.

of *Radix et Rhizoma Ginseng*. Compared to the ginsenoside Re reference peak (peak S), the relative retention times of the other 16 common peaks were 0.065, 0.070, 0.128, 0.360, 0.443, 0.475, 0.584, 0.860, 0.919, 1.099, 1.246, 1.318, 1.349, 1.390, 1.442 and 1.521.

Characteristics of Common Peaks

Peak 1 was fused with peak 2 and these two peaks were overlapping with multiple small peaks. Peak 4 was M-shaped. There were over 10 dentate peaks between peak 6 and peak 7. Peak 8 consisted of clustered peaks, which could be separated into over 15 small peaks, and its relative retention time was 0.740–0.860. There were 4–6 small peaks between peak 9 and peak S. Besides peak 10, there were over 10 dentate peaks between peak S and peak 11. There was usually a small peak separated before peak 12 and a small peak separated after peak 14.

Ratio of Common Peak Areas

The peak area of common peak 12 was more than 10% of the total

2.2

Daqingye
大青叶
Folium isatidis

Introduction

Folium isatidis is the dried leaf of *Isatis indigotica* Fort. (family Cruciferae). It mainly originates in Jiangsu, Anhui, Hebei and Henan Provinces.

According to some studies, *Folium isatidis* is planted abundantly in some areas of Anhui Province, such as Haozhou, Taihe, Linquan, Huaiyuan, Suxian, Lingbi, Mengcheng, Dangshan, Fuyang, Funan, Jieshou, Lixin, Guoyang, Sixian and Guzhen. It is sold not only within Anhui Province, but also in other provinces.

Chemical Composition

Nearly 1% of *Folium isatidis* is isatan B, which is easily hydrolyzed into indirubin and indigo. It also contains glucobrassicin, neoglucobrassicin, L-sulfo-glucobrassicin, sinigrin, free indanol and oxidase.

Experiment Materials

Samples

Samples of *Folium isatidis* were collected from one-year-old plants in Haozhou and Taihe (Anhui Province) in October 2002. There were 11 samples with variant identifications completed on all of them.

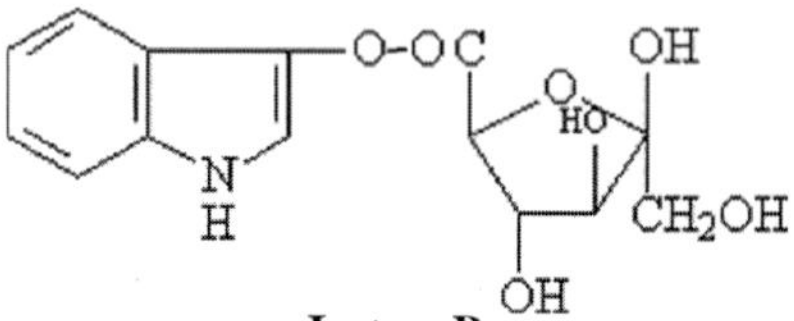

Isatan B

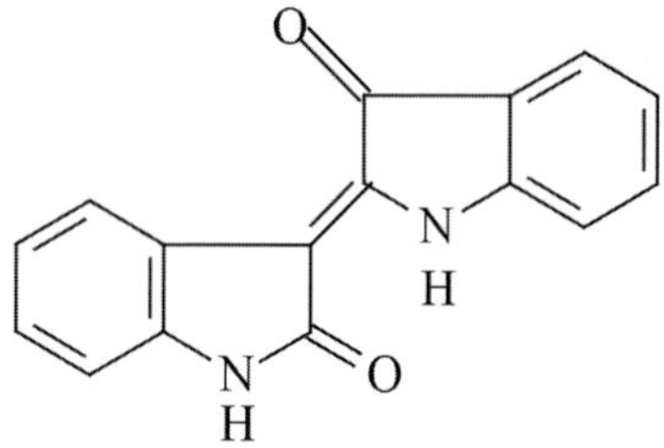

Indirubin

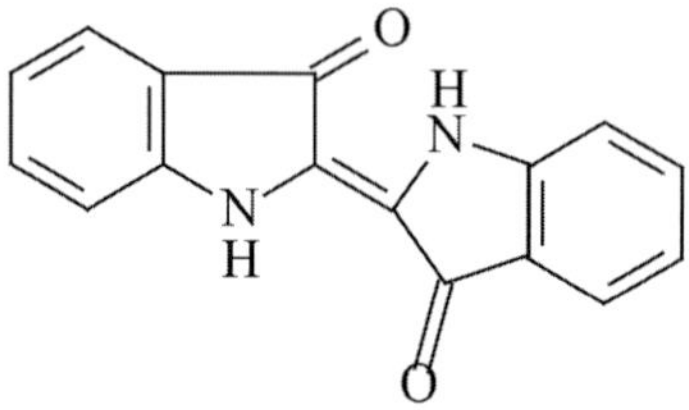

Indigo

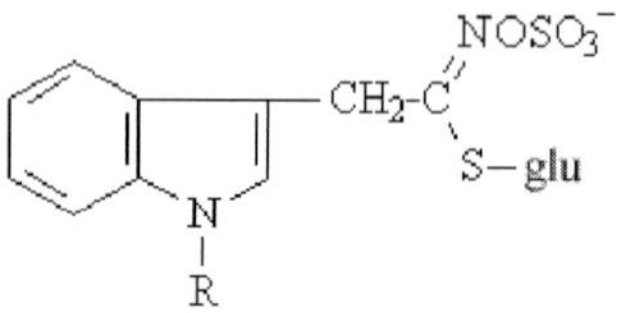

Glucobrassicin R=H
Neoglucobrassicin R=OCH_3
L-sulfo-glucobrassicin R=OSO_3^-

Reagents

Acetonitrile (chromatographic-grade, Tedia Company Inc., USA). Double-distilled deionized water (prepared by our lab). Methanol (analytical-grade methanol was purchased from Shanghai Chemical Reagents Pte. Ltd.; chromatographic-grade methanol was purchased from Tedia Company Inc.).

Preparation of Samples

2 g of crude powder of the sample was accurately weighed and placed in a round-bottom flask. 100 ml of methanol was added to the flask. After refluxing for 2 h, the sample solution was filtered. 100 ml of methanol was added to the residue and extracted for 1 h. After filtering, the two filtrates were combined. The filtrate was evaporated to dryness in an evaporating dish. The residue was dissolved and topped

up to 10 ml with methanol in a volumetric flask. The mixture was filtered through a 0.45 μm Millipore filter. The filtrate was then ready to be used as a test sample.

HPLC Conditions

Instrument: Agilent 1100 high-performance liquid chromatography and diode array detector
Mobile phase: Water and acetonitrile with gradient elution
Flow rate: 1.0 ml/min
Column: Agilent ZORBAX 80A Extend-C_{18} analytical column (250 mm × 4.6 mm, 5 μm)
Column temperature: 25°C
Detecting wavelength: 250 nm
Reference wavelength: 360 nm
Run time: 75 min
Injection volume: 20 μl

Methods

The sample solution was injected into the liquid chromatograph. The chromatogram was obtained under selected HPLC conditions. The experimental method was validated.

Mobile Phase Selection

A comparison of experimental results showed that gradient elution with water and acetonitrile could successfully separate the peaks and

A is water, B is acetonitrile.

	A (%)	B (%)
0–60 min	40–98	2–60
60–70 min	20–40	60–80
70–75 min	20–20	80–80

produce the maximum number of peaks from the samples. The duration of the analysis was 75 min. The 2 h chromatogram of the samples showed no characteristic peaks after 75 min.

Method Validation

1. Stability: Test samples from the same batch were stored at room temperature and injected at regular intervals for testing. The degrees of similarity of the injections calculated by the Similarity Degree Calculation Software for Traditional Chinese Medicine Fingerprints were all above 0.9, which indicated that samples were stable within the 36 testing hours.
2. Precision: Five consecutive injections of the samples from the same batch were conducted to determine the precision. The degrees of similarity of the injections calculated by the Similarity Degree Calculation Software for Traditional Chinese Medicine Fingerprints were all above 0.95, which indicated that the injections of the test samples were carried out with precision.
3. Reproducibility: Five samples from the same batch were prepared and tested according to the method described above. The degrees of similarity of the injections calculated by the Similarity Degree Calculation Software for Traditional Chinese Medicine Fingerprints were all above 0.9, indicating good reproducibility of the results.

Sample Analysis

The HPLC fingerprints of the test samples from 10 batches were analyzed under the conditions described above.

Establishment of Fingerprints

The number of peaks, peak area and relative retention time of 10 batches of samples were analyzed and compared to obtain an optimized fingerprint (Figs. 2.1, 2.2).

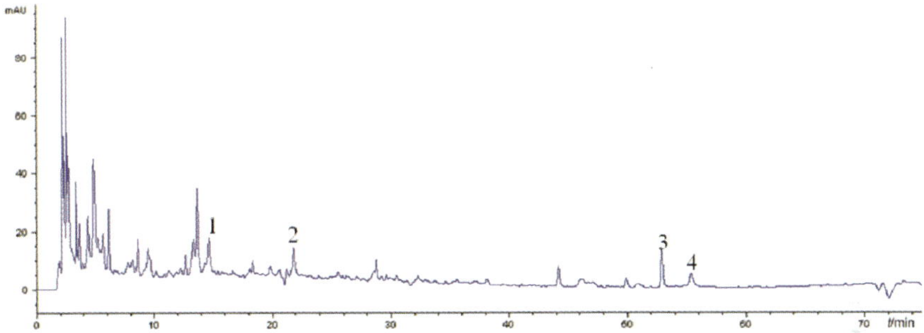

Fig. 2.1. HPLC fingerprint of Daqingye (*Folium isatidis*).

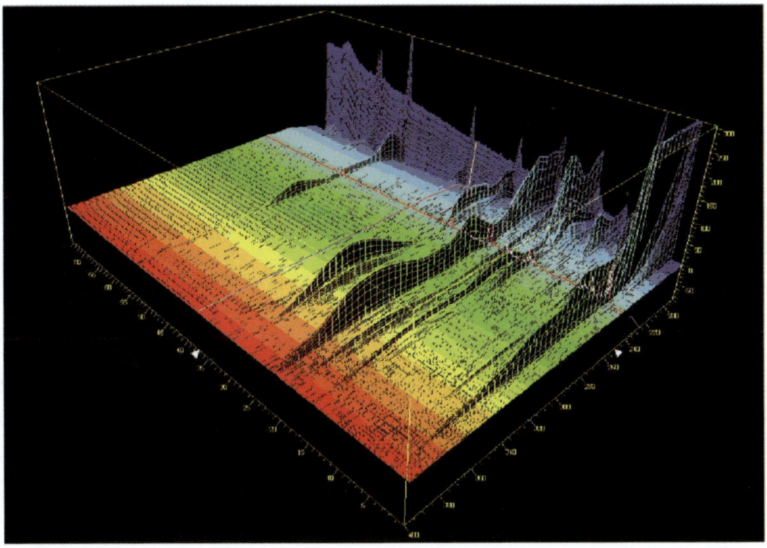

Fig. 2.2. 3-D HPLC fingerprint of Daqingye (*Folium isatidis*).

Results and Discussion

Over 30 peaks were separated by HPLC. After comparing all the tested and documented fingerprints, four common peaks were selected as the stable characteristic peaks of the *Folium isatidis* fingerprint. The retention time (min) and peak area values of the four common peaks are listed below:

Peak number	1	2	3	4
Retention time	14.641	21.792	52.854	55.326
Peak area	375.4	285.2	173.1	105.8

Common Peak Standardization

Common peaks were assigned according to the results of 10 batches of samples. There were four common peaks in the HPLC fingerprints of *Folium isatidis*. The common peak with a retention time of about 52.85 min was subsequently used as the reference peak. Compared to the reference peak (peak 3), the relative retention times of the other three common peaks were 0.277, 0.412 and 1.047.

Ratio of Common Peak Area

None of the peak areas of the common peaks exceeded 10% of the total peak area.

2.3

Dahuang
大黄
Radix et Rhizoma Rhei

Introduction

Radix et Rhizoma Rhei (family Polygonaceae) is the dried root or rhizome of *Rheum palmatum* L., *Rheum tanguticum* Maxim. ex Balf. or *Rheum officinale* Baill.

Rheum palmatum L. and *Rheum tanguticum* Maxim. ex Balf. are mainly produced in Gansu, Qinghai, Xizhang and Sichuan Provinces. Both are called Western Dahuang or Northern Dahuang. Medicinal Dahuang (*Rheum officinale* Baill.), which is also called Southern Dahuang or Chuan Dahuang, is mainly produced in Sichuan, Guizhou, Yunnan, Hubei and Shaanxi. The output of *Rheum palmatum* L. is higher than that of *Rheum tanguticum*. Medicinal Dahuang (*Rheum officinale* Baill.) is rarely produced.

According to studies on resources, the cultivated and wild *Rheum palmatum* L. are distributed in Minxian, Wenxian, Lixian, Linxia, Zhouqu, Diebu, Danni, Luqu, Yanchang, Tianzhu, Huating, Zhuanglang, Wudou, Yongdeng, Xiahe and Hezheng in Gansu Province. It is produced abundantly in Minxian, Lixian, Wenxian and Zhuanglang, which have a long history of cultivating this plant. The planting area of *Radix et Rhizoma Rhei* in Zhuanglang and Lixian is about 6666.67 hm^2 with an abundant quantity of about 50,000 tons produced per year.

Chemical Composition

Radix et Rhizoma Rhei contains anthraquinones, such as rhein, emodin, chrysophanol, aloe-emodin, physcion, sennosides A, B, C, D, E, F, rhein-8-mono-β-D-glucoside, physcion monoglucoside, aloe-emodin monoglucoside, emodin monoglucoside, and chrysophanol-1-monoglucoside. It also contains butylphenosides, tannin and related polyphenols. The chemical structures of some anthraquinones and gallic acids are listed below.

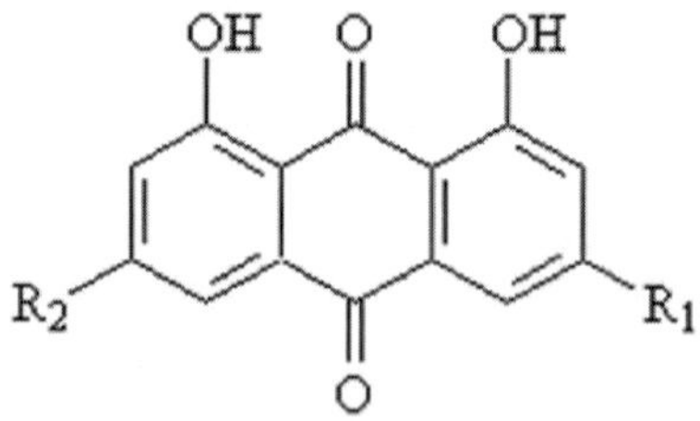

	R_1	R_2
Emodin	**CH_3**	**OH**
Chrysophanol	**CH_3**	**H**
Aloe-emodin	**CH_2OH**	**H**
Rhein	**COOH**	**H**
Physcion	**CH_3**	**OCH_3**

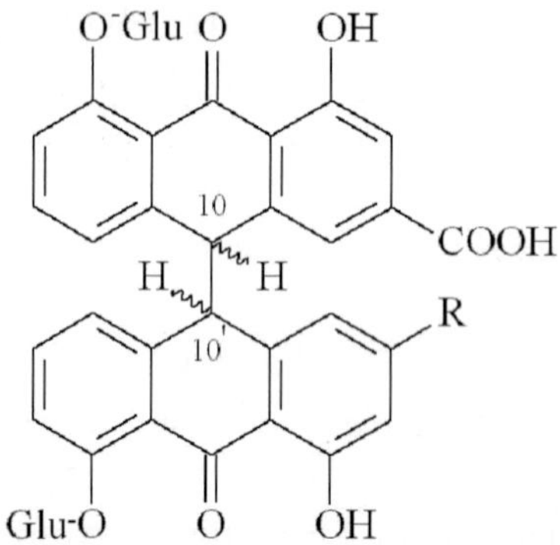

	R	10-10'
Sennoside A	**COOH**	***trans***
Sennoside B	**COOH**	***cis***
Sennoside C	**CH_2OH**	***trans***
Sennoside D	**CH_2OH**	***cis***

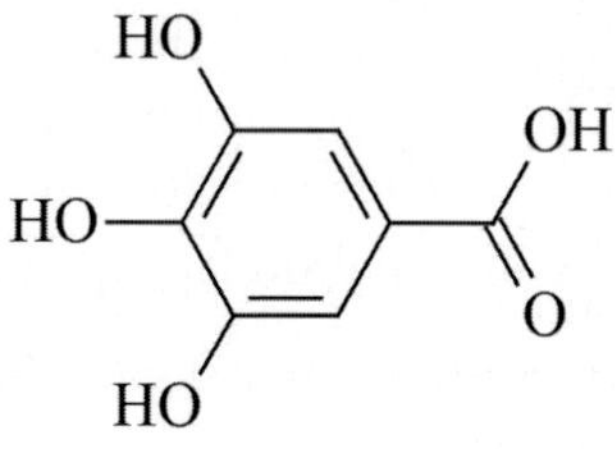

Gallic acid

Experiment Materials

Samples

Radix et Rhizoma Rhei samples were collected from three-year-old cultivated plants in Huating, Zhuanglang, Lixian and Dangchang in Ganshu Province in November 2001. There were 15 samples identified as *Rheum palmatum* L.

References

The purities of gallic acid and emodin, which were purchased from National Institute for the Control of Pharmaceutical and Biological Products (China), were 98.14% and 99.02% respectively. They were used as references for quantitative analysis.

Reagents

Acetonitrile (chromatographic-grade, Tedia Company Inc., USA). Methanol (analytical-grade methanol was purchased from Shanghai Chemical Reagents Pte. Ltd.; chromatographic-grade methanol was purchased from Tedia Company Inc.). Acetic acid (analytical-grade, Nanjing First Chemical Reagent Plant). Double-distilled deionized water (freshly prepared by our lab).

Preparation of Samples

2 g of powder sample was accurately weighed and passed through a 20 mesh sieve. 50 ml of methanol was added. After being refluxed in a water bath at 80°C for 1 h, the sample solution was filtered. 50 ml of methanol was added to the residue. The solution was refluxed in a water bath at 80°C for 1 h and was filtered. Both filtrates were combined and evaporated partially in an evaporating dish using a water bath. It was then topped up to 25 ml with methanol. The solution was filtered through a 0.45 μm Millipore filter. The filtrate was then ready to be used as a test sample.

Preparation of Reference

5 mg of gallic acid and 5 mg of emodin were accurately weighed, dissolved and topped up to 25 ml with methanol in a volumetric flask. The concentration of the solution was 0.2 mg/ml. After being diluted 10 times, it was then ready to be used as a reference.

HPLC Conditions

Instrument: Agilent 1100 high-performance liquid chromatography and diode array detector
Mobile phase: 1% HAc in water and 1% HAc in acetonitrile with gradient elution
Flow rate: 1.0 ml/min
Column: Inertsil® ODS analytical column (250 mm × 4.6 mm, 5 μm)
Column temperature: 30°C
Detecting wavelength: 280 nm
Reference wavelength: 380 nm
Run time: 90 min
Injection volume: 20 μl

Methods

The sample solution was injected into the liquid chromatograph. The chromatogram was obtained under selected HPLC conditions. The experimental method was validated.

Mobile Phase Selection

A comparison study showed that gradient elution using 1% HAc in water and 1% HAc in acetonitrile could successfully separate the chromatographic peaks with the maximum number of peaks. The duration of the analysis was 90 min. The 2 h chromatogram showed no characteristic peaks after 90 min.

A is 1% HAc in water, B is 1% HAc in acetonitrile.

	A (%)	B (%)
0–55 min	75–100	0–25
55–75 min	45–75	25–55
75–90 min	15–45	55–85

Method Validation

1. Stability: Test samples from the same batch (batch 9: Dahuang from Zhuanglang) were analyzed at 0 h, 2 h, 4 h, 7 h, 11 h, 16 h, 22 h and 36 h. The degrees of similarity of the injections calculated by the Similarity Degree Calculation Software for Traditional Chinese Medicine Fingerprints were all above 0.92, which indicated that samples were stable within the 36 testing hours.
2. Precision: Five consecutive injections of the samples from the same batch (batch 9: Dahuang from Zhuanglang) were conducted to determine the precision. The degrees of similarity of the injections calculated by the Similarity Degree Calculation Software for Traditional Chinese Medicine Fingerprints were all above 0.96, which indicated that the instrument is of acceptable precision.
3. Reproducibility: Five samples from batch 9 were prepared and analyzed according to the method described above. The degrees of similarity of the injections calculated by the Similarity Degree Calculation Software for Traditional Chinese Medicine Fingerprints were all above 0.9, indicating good reproducibility of the results.

Sample Analysis

The HPLC fingerprints of the test samples from 10 batches were analyzed under the conditions described above.

Establishment of Fingerprints

The number of peaks, peak area and relative retention time of 10 batches of samples were analyzed and compared to obtain an optimized fingerprint (Figs. 3.1, 3.2).

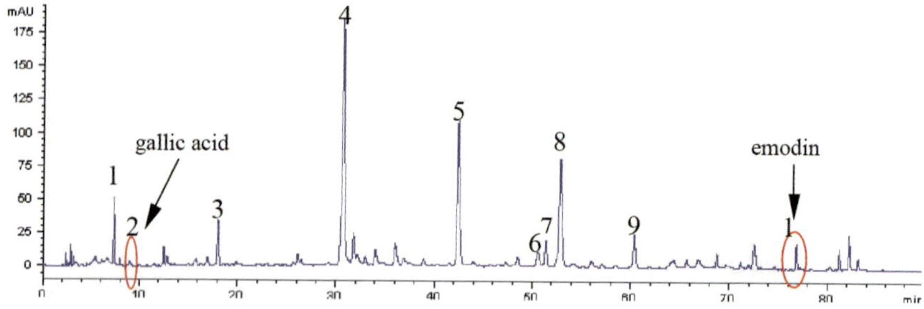

Fig. 3.1. HPLC fingerprint of Dahuang (*Radix et Rhizoma Rhei*).

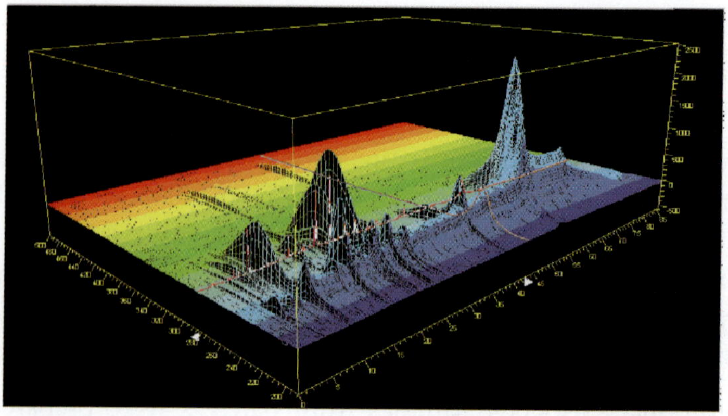

Fig. 3.2. 3D HPLC fingerprint of Dahuang (*Radix et Rhizoma Rhei*).

Results and Discussion

About 90 peaks were separated by HPLC. After comparing all the chromatograms, 10 common peaks were selected as the characteristic peaks. The retention time (min) and peak area values of the 10 common peaks are listed below.

Peak number	1	2	3	4	5	6	7	8	9	10
Retention time	7.25	8.945	18.04	30.77	42.54	51.61	52.65	53.06	60.71	76.44
Peak area	965.14	207.61	898.25	5824.05	8354.54	972.54	201.89	2852.43	1336.58	29.68

Common Peak Standardization

Common peaks were assigned according to the results of 10 batches of samples. There were 10 common peaks in the HPLC fingerprints of *Radix et Rhizoma Rhei*. Compared to the reference peak of emodin (peak 10), the relative retention times of the other nine common peaks were 0.095, 0.117, 0.236, 0.403, 0.557, 0.675, 0.689, 0.694 and 0.794.

Ratio of Common Peak Area

The peak areas of both peaks 4 and 5 exceeded 10% the total peak area. The ratio of peak 4 to the reference peak was specified to be within 137.360–255.096, whilst that for peak 5 was specified to be within 197.041–365.933.

Evaluation of Similarity

The data obtained from the chromatogram was input into the Similarity Degree Calculation Software for Traditional Chinese Medicine Fingerprints. The peaks were selected and matched automatically with the matching template. Then the standard template was setup and the differences among the peaks and the overall similarity were evaluated. Figure 3.3 shows the overlapping pattern of the

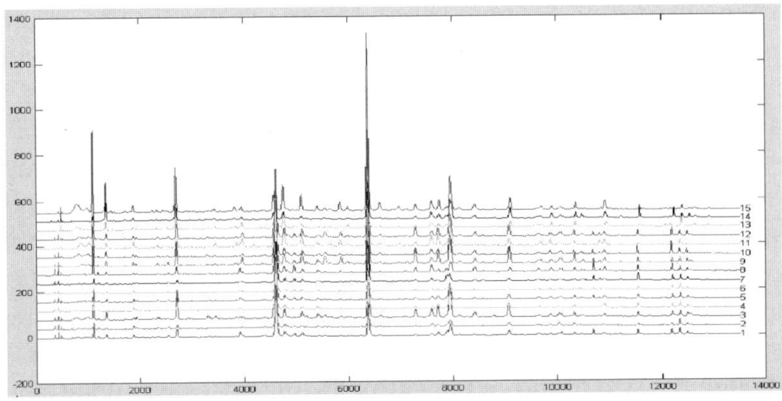

Fig. 3.3. Overlapping pattern of HPLC fingerprints of *Rheum palmatum* L.

HPLC fingerprints of *Rheum palmatum* L. Samples from Huating, Zhuanglang, Lixian and Dangchang were highly similar and the similarities of most of the samples were within the range of 0.85–0.95. It indicated that samples from the above four locations were highly correlated and yet different.

Analysis of Principal Component

The singular values of fingerprint data matrix are factorized into two orthogonal matrixes (**U**, **V**) and a diagonal matrix (**S**). The product of two orthogonal matrixes (**U**, **V**) multiplied by the diagonal matrix (**S**) is then calculated. Two of the three values are used as the ordinate and abscissa in the projection diagram. The other value is the projection point in the projection diagram. The changing patterns of the principal components can be identified according to the different places the projection points are located. The *Radix et Rhizoma Rhei* samples can be classified into two types. The first type was different from the main chromatogram but the other one was similar to it (of which samples 13 and 14 were from Dangchang, but samples 11 and 15 were collected in 2000). In order to further understand the differences between them, chromatograms of samples 1 (from Huating), 10, 11, 13, 14 and 15 were selected and their HPLC fingerprints were overlapped in one chromatogram. The six individual fingerprints showed that the differences were defined by the size of the peak with a retention time of 42 min. This is because this area of this peak is bigger than that of the others, which indicates that it has a higher weight coefficent. Hence, the size of this peak would affect the extracted quantity of the whole medicinal material. Comparison of the six chromatograms showed a gradual increase of this peak in samples 1, 14, 13, 10, 11 and 15. The biggest area of this peak was found in samples 11 and 15, which might be due to chemical degradation that occurred during the prolonged storage. The chemical degradation from other substances caused this peak to increase in size and caused the deviation from the main chromatogram in samples 11 and 15. Although the size of this peak in samples 14 and 13 was quite similar to that in sample 1, the overall degree of similarity was much different from that of sample 1, so that samples 14 and 13 deviated tremendously from the main chromatogram. This was because of the effects of other ingredients in the samples. Samples 1 and 10 had a high degree of similarity. However, the size of peaks with a retention time of 42 min was different. Hence, there is a difference in the projected graph. It is worth mentioning that samples with a lower total quantity are located to the left end of the main component projection diagram. As the position of the sample point is shifted from the left to the right part of the principal component projection diagram, the total peak area in the whole fingerprint becomes bigger and bigger.

2.4

Shanzhuyu
山茱萸
Fructus corni

Introduction

Fructus corni (family Cornaceae) is the dried fruit of *Cornus officinalis* Seib. et Zucc.

Fructus corni mainly originates in Zhejiang Province. According to studies on resources, it is cultivated abundantly in Nanyang Xixia, Neixiang, Nanzhao, Zhechuan and Zhenping in Henan Province.

Chemical Composition

Fructus corni mainly contains glycosides, tannins and volatile oils. Glycosides include cornin, morroniside, 7-*O*-methyl morroniside, sweroside, loganin and cornuside. Its aglucons are ursolic acid and iridoids. Tannins include cornus-tannins 1, 2 and 3. The chemical structures of some glycosides and aglucons are listed below.

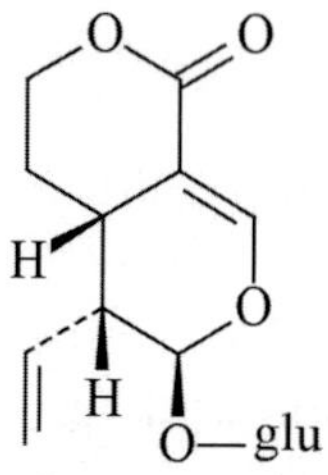

Sweroside

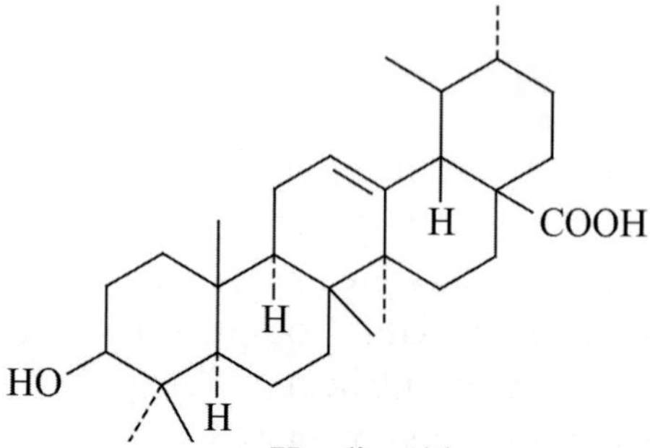

Ursolic acid

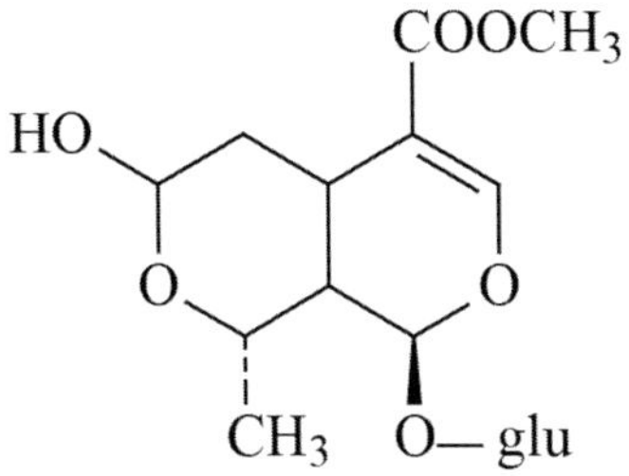

Morroniside

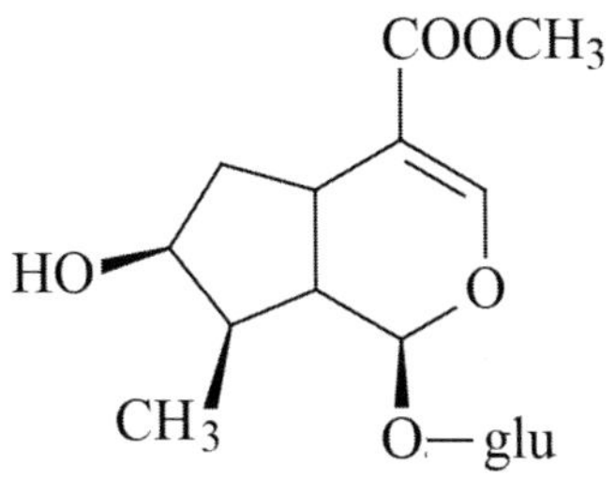

Loganin

Experiment Materials

Samples

Cultivated *Fructus Corni* samples were collected from Nanyang Xixia, Neixiang, Nanzhao, Zhechuan and Zhenping in Henan Province. There were 15 samples with variant identifications completed on all of them.

References

Loganin and ursolic acid used for quantitative analysis were provided by the National Institute for the Control of Pharmaceutical and Biological Products (China). Morroniside and sweroside were self-prepared.

Reagents

Acetonitrile (chromatographic-grade, Tedia Company Inc., USA). Methanol (analytical-grade methanol was purchased from Shanghai Chemical Reagents Pte. Ltd.; chromatographic-grade methanol was purchased from Tedia Company Inc.). Phosphoric acid (analytical-grade, Nanjing First Chemical Reagent Plant). Double-distilled deionized water (prepared by our lab).

Preparation of Samples

0.5 g of powder sample was accurately weighed and passed through a 40 mesh sieve. 20 ml of methanol was added. After being refluxed in a water bath for 1 h, the sample solution was filtered. The residue was refluxed and filtered using the above method twice. All the three filtrates were combined and the solvent was evaporated off at low temperature. The residue was dissolved in methanol, topped up to 25 ml and mixed well. The solution was then filtered through a 0.45 μm Millipore filter and the first 5 ml of filtrate was discarded. The filtrate was then ready to be used as a test sample.

Preparation of Reference

Morroniside, loganin, sweroside and ursolic acid were accurately weighed and dissolved in methanol to obtain a concentration of 50 μg/ml respectively for analysis.

HPLC Conditions

Instrument: Agilent 1100 high-performance liquid chromatography and diode array detector
Mobile phase: 0.1% phosphoric acid in water and 0.1% phosphoric acid in acetonitrile with gradient elution
Flow rate: 1.0 ml/min
Column: Phenomenex® Luna C_{18} analytical column (250 mm × 4.6 mm, 5 μm)
Column temperature: 30°C
Detecting wavelength: 240 nm
Reference wavelength: 360 nm
Run time: 70 min
Injection volume: 20 μl

Methods

The sample solution was injected into the liquid chromatograph to obtain a chromatogram under selected HPLC conditions. The experimental method was validated.

Mobile Phase Selection

A comparison study showed that gradient elution using 0.1% phosphoric acid in water and 0.1% phosphoric acid in acetonitrile with gradient elution could successfully separate the chromatographic peaks with the maximum number of peaks. The duration of the analysis was 70 min. The 2 h chromatogram showed no characteristic peaks after 70 min.

A is 0.1% H_3PO_4 in water, B is 0.1% H_3PO_4 in acetonitrile.

	A (%)	B (%)
0–10 min	93–93	7–7
10–45 min	80–93	7–20
45–65 min	80–80	20–20

Method Validation

1. Stability: Samples from the same batch were analyzed at 5 h, 7 h, 8 h, 9 h and 10 h. The degrees of similarity of the injections calculated by the Similarity Degree Calculation Software for Traditional Chinese Medicine Fingerprints were all above 0.99, which indicated that samples were stable within the 36 testing hours.
2. Precision: Five consecutive injections of the samples from the same batch were conducted to determine the precision. The degrees of similarity of the injections calculated by the Similarity Degree Calculation Software for Traditional Chinese Medicine Fingerprints were all 1, which indicated that the instrument is of acceptable precision.

3. Reproducibility: Five samples from the same batch were prepared and analyzed according to the method described above. The degrees of similarity of the injections calculated by the Similarity Degree Calculation Software for Traditional Chinese Medicine Fingerprints were all above 0.99, indicating good reproducibility of the results.

Sample Analysis

The HPLC fingerprints of the test samples from 10 batches were analyzed under the conditions described above.

Establishment of Fingerprints

The number of peaks, peak area and relative retention time of 10 batches of samples were analyzed and compared to obtain an optimized fingerprint (Figs. 4.1, 4.2).

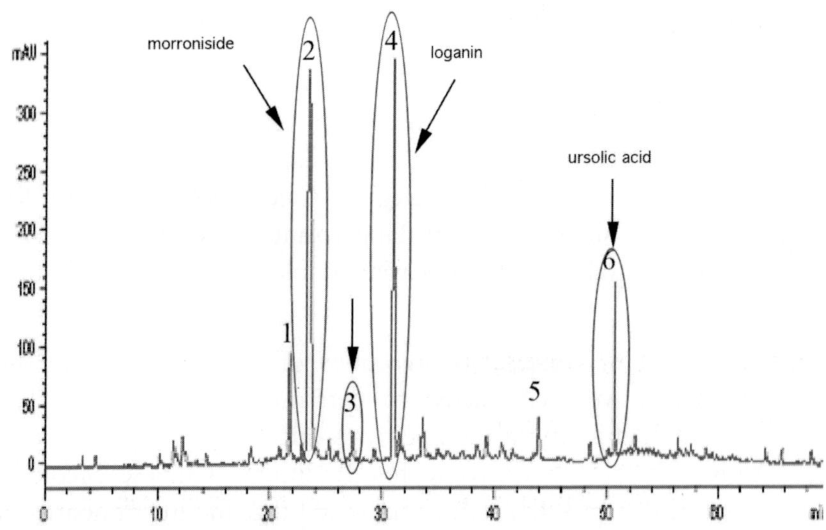

Fig. 4.1. HPLC fingerprint of Shanzhuyu (*Fructus corni*).

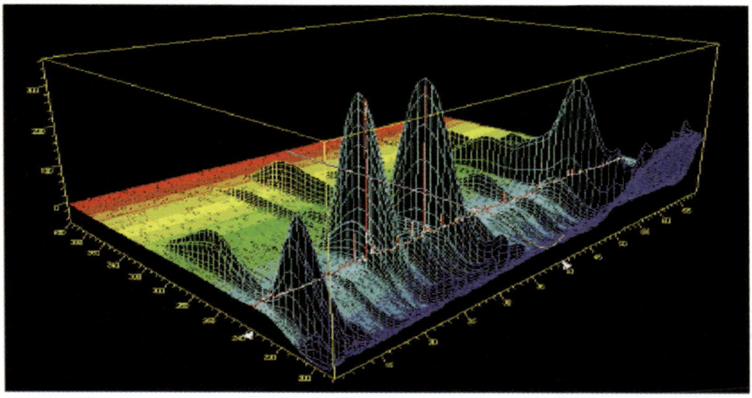

Fig. 4.2. 3-D HPLC fingerprint of Shanzhuyu (*Fructus corni*).

Results and Discussion

About 50 peaks were separated by HPLC. After comparing them with the reference, six common peaks were selected as the characteristic peaks. The retention time (min) and peak area values of the 10 common peaks are listed below.

Peak number	1	2	3	4	5	6
Retention time	21.68	23.39	29.23	30.93	43.9	50.64
Peak area	356.19	3768.77	66.51	1749.32	92.6	364.4

Common Peak Standardization

Common peaks were assigned according to the results of 10 batches of samples. There were six common peaks in the HPLC fingerprints of *Fructus corni*. Compared to the loganin reference peak (peak 4), the relative retention times of the other five common peaks were 0.701, 0.756, 0.945, 1.419 and 1.637.

Ratio of Common Peak Area

Aside from the loganin reference peak, the area of peak 2 (sweroside) was bigger than 10% of the total peak area. The ratio of peak 2 to the reference peak was specified to be within 1.508–2.800.

2.5

Shanyao
山药
Rhizoma dioscoreae

Introduction

Rhizoma dioscoreae (family Dioscoreaceae) is the dried rhizome of *Dioscorea opposita* Thunb.

Rhizoma dioscoreae mainly originates from Henan Province. There are two types of commercially available Shanyao: one is Guang Shanyao (光山药) and the other is Mao Shanyao (毛山药). According to studies on resources, it is planted abundantly in Wenxian, Wuzhi, Bo'ai, Qinyang and Mengxian of Henan Province. The *Rhizoma dioscoreae* cultivated in those areas is of high quality, which ensures steady sales not only in China, but also in other countries.

Chemical Composition

Rhizoma dioscoreae contains batatasins I, II, III, IV, V, allantoin, choline, arginine, glucoprotein and mannan.

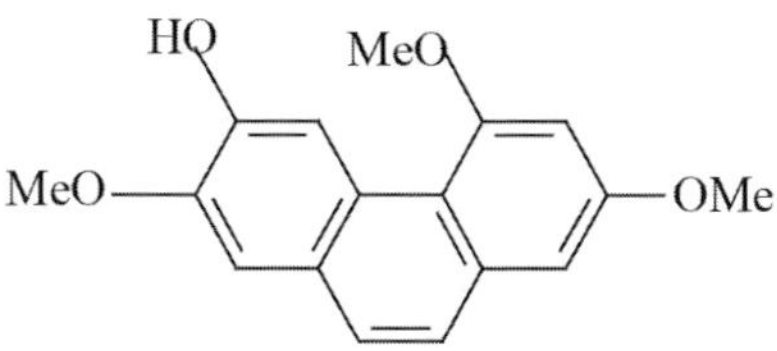

Batatasin I

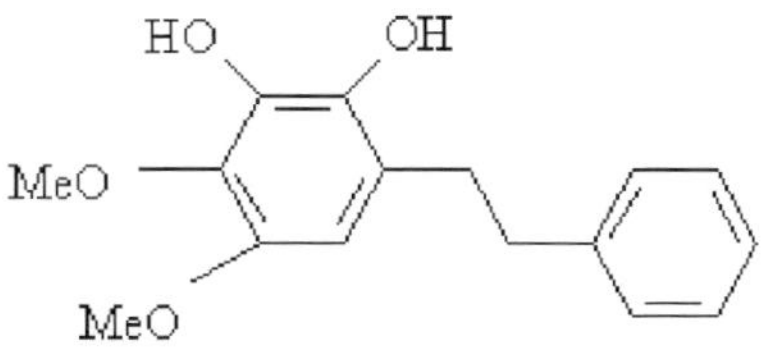

Batatasin II

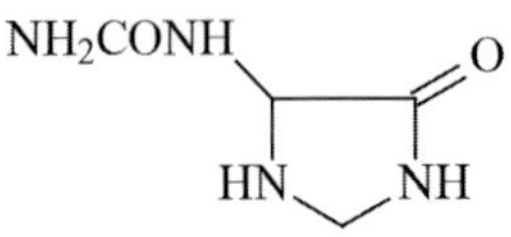

Allantoin

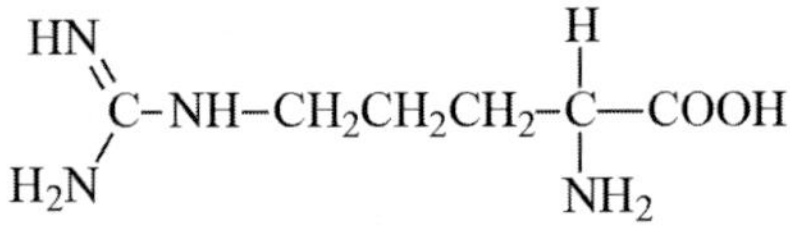

Arginine

Experiment Materials

Samples

Rhizoma dioscoreae samples were collected from Wude and Yuecun of Wenxian, Xitao, Dafeng of Wuzhixian, etc, in Jiaozuo of Henan Province in October 2002. There were 15 samples identified as *Rhizoma dioscoreae*.

Reference

Allantoin (lot number 1501-200001) was used for quantitative analysis, and was provided by the National Institute for the Control of Pharmaceutical and Biological Products (China).

Reagents

Methanol (chromatographic-grade, Tedia Company Inc., USA). Phosphoric acid (analytical-grade, Nanjing First Chemical Reagent Plant). Double-distilled deionized water (prepared by our lab).

Preparation of Samples

After grinding, 10 g of powder sample (passing through an 80 mesh sieve) was accurately weighed and placed in a 250 ml volumetric flask. 100 ml of fresh double-distilled H_2O was accurately transferred into the flask. The solution was sonicated with an operating frequency of 40 kHz and power of 500 W for 20 min. The sonication was repeated for two more times. The extract was centrifuged at high speed (10,000 r/min) and filtered through a 0.45 μm Millipore filter. The filtrate was then ready to be used as a test sample.

Preparation of Reference

5 mg of allantoin was accurately weighed, dissolved and topped up to 50 ml with methanol in a volumetric flask to obtain a concentration of 0.1 mg/ml. It was then ready to be used as a reference.

HPLC Conditions

Instrument: Agilent 1100 high-performance liquid chromatography and diode array detector
Mobile phase: 0.1% H_3PO_4 in water and methanol with gradient elution
Flow rate: 1.0 ml/min
Column: Alltech ODS C_{18} (250 mm × 4.6 mm, 5 μm)
Column temperature: 30°C
Detecting wavelength: 258 nm
Reference wavelength: 360 nm
Run time: 40 min
Injection volume: 20 μl

Methods

The sample solution was injected into the liquid chromatograph. The chromatogram was obtained under selected HPLC conditions. The experimental method was validated.

Mobile Phase Selection

A comparison study showed that gradient elution using 0.1% H_3PO_4 in water (A) and methanol (B) could successfully separate the chromatographic peaks with the maximum number of peaks (0–40 min, A: 90–99.5%, B: 0.5–10%). The duration of the analysis was 40 min. The 1 h chromatogram showed no characteristic peaks after 40 min.

Method Validation

1. Stability: Samples from the same batch were stored at room temperature and injected at regular intervals. The degrees of similarity of the injections calculated by the Similarity Degree Calculation Software for Traditional Chinese Medicine Fingerprints were all above 0.93, which indicated that samples were stable within the 36 testing hours.
2. Precision: Five consecutive injections of the samples from the same batch were conducted to determine the precision. The degrees of similarity of the injections calculated by the Similarity Degree Calculation Software for Traditional Chinese Medicine Fingerprints were all above 0.97, which indicated that the instrument is of acceptable precision.
3. Reproducibility: Five samples from the same batch were prepared and analyzed according to the method described above. The degrees of similarity of the injections calculated by the Similarity Degree Calculation Software for Traditional Chinese Medicine Fingerprints were all above 0.94, indicating good reproducibility of the results.

Sample Analysis

The HPLC fingerprints of the test samples from 10 batches were analyzed under the conditions described above.

Establishment of Fingerprints

The number of peaks, peak area and relative retention time of 10 batches of samples were analyzed and compared to obtain an optimized fingerprint (Fig. 5.1).

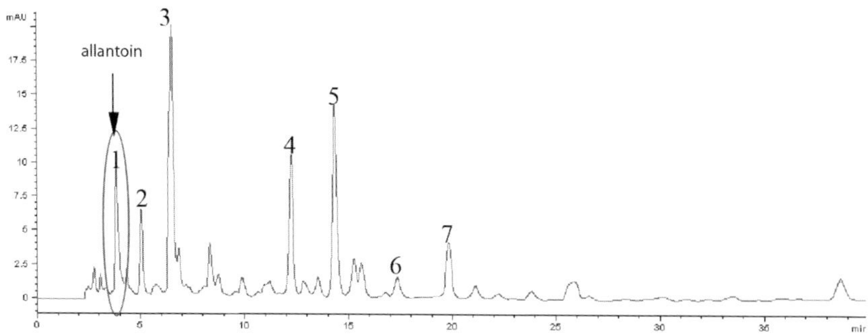

Fig. 5.1. HPLC fingerprint of Shanyao (*Rhizoma dioscoreae*).

Results and Discussion

Over 30 peaks were separated by HPLC. After comparing all the chromatograms with the reference, seven common peaks were selected as the characteristic peaks. The retention time (min) and peak area values of the 10 common peaks are listed below.

Peak number	1	2	3	4	5	6	7
Retention time	3.809	5.035	6.52	12.307	14.414	17.387	19.985
Peak area	101	60.8	278.1	126.5	206	23.1	68.2

Common Peak Standardization

Common peaks were assigned according to the results of 10 batches of samples. There were seven common peaks in the HPLC fingerprints of *Rhizoma dioscoreae*. Compared to the reference peak of allantoin (peak 1), the relative retention times of the other six common peaks were 1.322, 1.712, 3.231, 3.784, 4.565 and 5.247.

Ratio of Common Peak Area

The peak areas of peaks 3, 4 and 5 exceeded 10% of the total peak area. The ratio of peak 3 to the reference peak was specified to be within 1.927–3.579, whilst the ratios for peaks 4 and 5 were specified to be within 0.876–1.628 and 1.428–2.652 respectively.

2.6

Chuanxiong
川芎
Rhizoma chuanxiong

Introduction

Rhizoma chuanxiong (family Umbelliferae) is the dried rhizome of *Ligusticum chuanxiong* Hort.

Rhizoma chuanxiong is mainly cultivated in Sichuan Province. According to studies on resources, Dujiangyan (Guanxian, Sichuan) is famous for the greatest abundance of *Rhizoma chuanxiong*. The *Rhizoma chuanxiong* produced in Chongzhou, Pengzhou, Dujiangyan, Pixian, Xindu and Cangxi, cultivated using the "Exchanged Location for Planting" method, is of large size and good quality.

Chemical Composition

Rhizoma chuanxiong contains volatile oils, phthalides and alkaloids. Phthalides include 4-hydroxy-3-butylphthalide, ligustilide, butylidenephthalide, butylphthalide, neocnidilide, cnidilide and senkyunolide A. There are other ingredients as well, such as tetramethylpyrazine, chrysophanol, sedanonic acid and 1-(5-hydroxymethyl-2-furanyl)-9H-pyridine(3,4,6) indole. Some of the chemical structures are listed below.

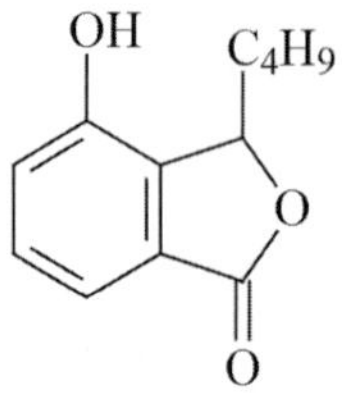

4-Hydroxy-3-butylphthalide

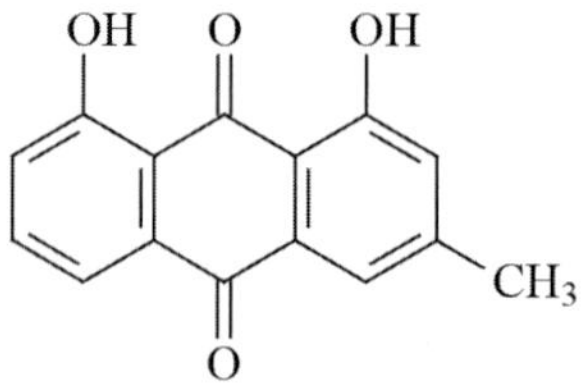

Chrysophanol

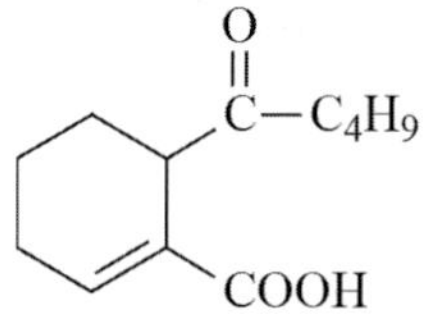

Sedanonic acid

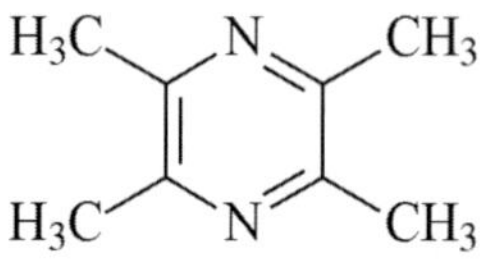

Chuanxiongzine (Tetramethylpyrazine)

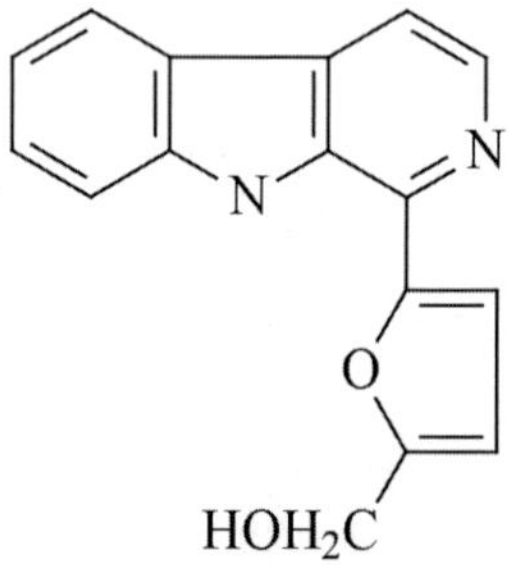

1-(5-Hydroxymethyl-2-furanyl) -9H-pyridine(3,4,6) indole

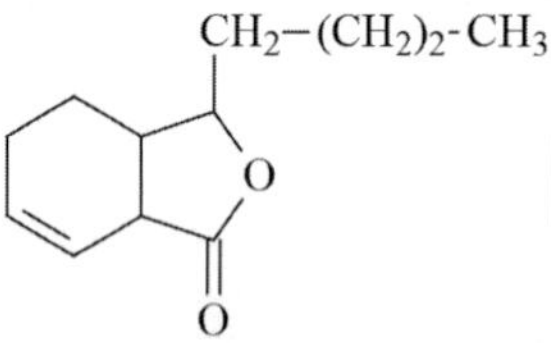

Cnidilide

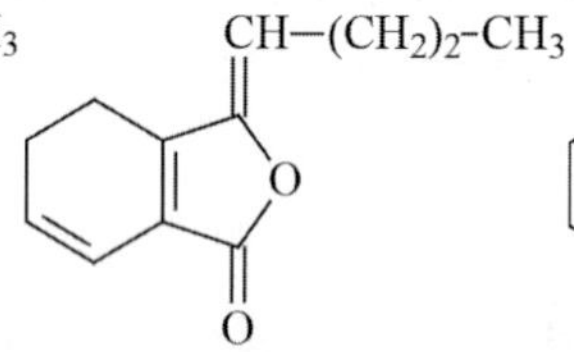

Ligustilide

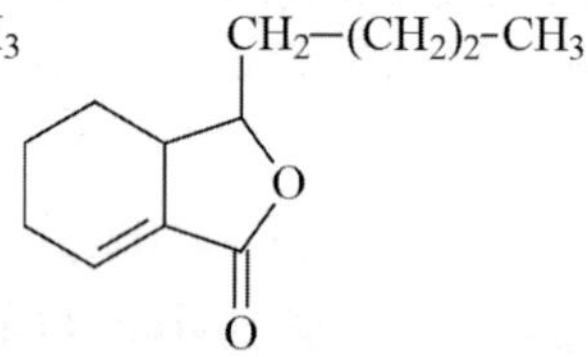

Neocnidilide

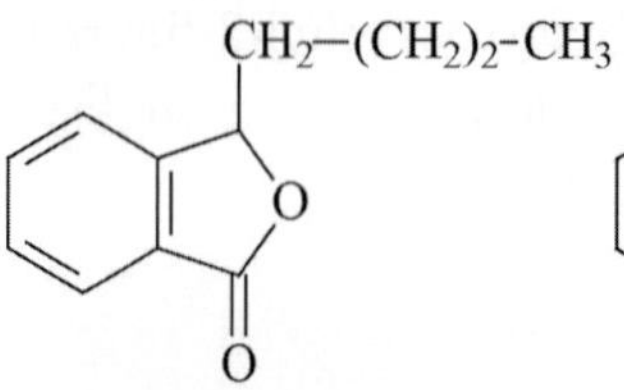

Butylphthalide

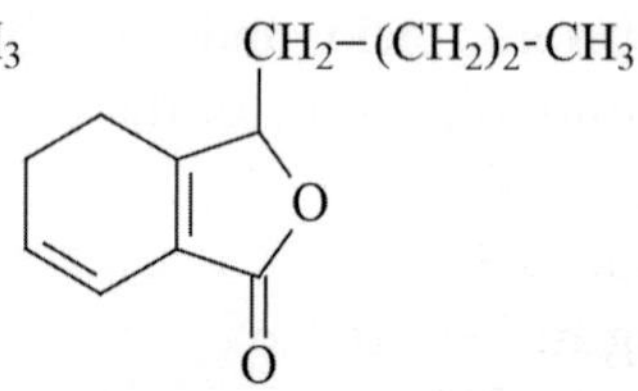

Senkyunolide

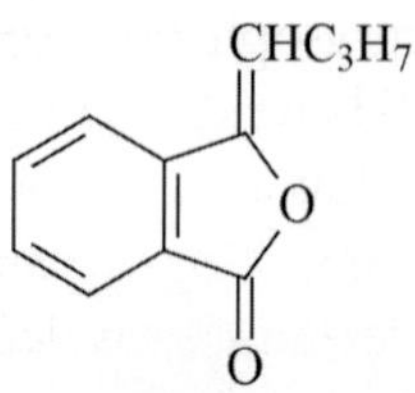

Butylidene phthalide

Experiment Materials

Samples

Rhizoma chuanxiong samples were collected from the cultivation bases in Pengzhou, Dujiangyan, Pixian, Chongzhou and Xinfan of Sichuan Province. There were 12 samples identified as *Rhizoma chuanxiong*.

Reagents

Methanol (chromatographic-grade, Tedia Company Inc., USA). 95% ethanol and phosphoric acid (analytical-grade). Double-distilled deionized water (prepared by our lab).

Preparation of Samples

2 g of sliced raw samples were accurately weighed and placed in a round-bottom flask. 30 ml of methanol was added to the flask and the sample was refluxed in a water bath at 70°C. After boiling for 1 h, the mixture was filtered. The residue was refluxed again and filtered using the above steps. The two filtrates were combined, concentrated and topped up to 10 ml with methanol. The solution was filtered with a 0.45 Millipore filter and was then ready to be used as a test sample.

HPLC Conditions

Instrument: Agilent 1100 high-performance liquid chromatography and diode array detector
Mobile phase: 0.5% H_3PO_4 in water and 0.5% H_3PO_4 in methanol with gradient elution
Flow rate: 1.0 ml/min
Column: Kromasil KR100-5 C_{18} analytical column (150 mm × 4.6 mm)
Column temperature: 25°C
Detecting wavelength: 280 nm
Reference wavelength: 360 nm

Run time: 60 min
Injection volume: 10 μl

Methods

The sample solution was injected into the liquid chromatograph. The chromatogram was obtained under selected HPLC conditions. The experimental method was validated.

Mobile Phase Selection

A comparison study showed that gradient elution using 0.5% H_3PO_4 in water (A) and 0.5% H_3PO_4 methanol (B) could successfully separate the chromatographic peaks with the maximum number of peaks. The duration of the analysis was 60 min. The 1.5 h chromatogram showed no characteristic peaks after 60 min.

A is 0.5% H_3PO_4 in water and B is 0.5% H_3PO_4 methanol.

	A (%)	B (%)
0–3 min	85–85	15–15
3–55 min	5–85	15–95
55–60 min	5–5	95–95

Method Validation

1. Stability: Samples from the same batch were analyzed at 0 h, 1 h, 2 h, 4 h, 8 h and 12 h to check the consistency of the relative retention time of each injection. The relative retention time and RSD of the peak area of the main peaks were both lower than 3%. They were within the requirements and indicated that samples were stable enough.
2. Precision: Six consecutive injections of the samples from the same

batch were conducted to determine the precision. The RSD values of the relative retention times of the main peaks were all lower than 1.5%. They were within the requirements and indicated that injections of samples were precise enough.

3. Reproducibility: Five samples from the same batch were prepared and analyzed according to the method described above. The RSD values of the relative retention times of all peaks were all lower than 3%, indicating good reproducibility of the method.

Sample Analysis

The HPLC fingerprints of the test samples from 10 batches were analyzed under the conditions described above.

Establishment of Fingerprints

The number of peaks, peak area and relative retention time of 10 batches of samples were analyzed and compared to obtain an optimized fingerprint (Fig. 6.1).

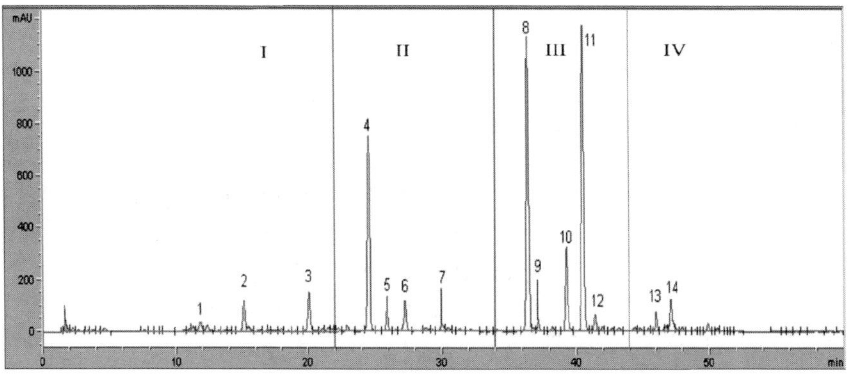

Fig. 6.1. HPLC fingerprint of Chuanxiong (*Rhizoma chuanxiong*).

Results and Discussion

About 100 peaks were separated by HPLC. After comparing all the chromatograms, 14 common peaks were selected as the characteristic peaks. In order to easily identify them, each chromatogram was divided into four zones according to the retention time. The first zone, which contains peaks 1 to 3, lies between 10 and 22 min. The second zone, which contains peaks 4 to 7, lies between 22 and 34 min. The third zone, which contains peaks 8 to 12, lies between 34 and 44 min. The fourth zone, which contains peaks 13 to 14, lies between 44 and 54 min.

Common Peak Standardization

Common peaks were assigned according to the results of 10 batches of samples. There were 14 common peaks in the HPLC fingerprints of *Rhizoma chuanxiong*. The common peak with a retention time of about 36.4 min was subsequently used as the reference peak. Compared to the reference peak (peak 8), the relative retention times of the other 13 common peaks were 0.3234, 0.4132, 0.5501, 0.6736, 0.7114, 0.7504, 0.8247, 1.0221, 1.0813, 1.1145, 1.1388, 1.2653 and 1.2947.

Ratio of Common Peak Area

Other than the reference peak (peak 8), the area of peak 11 was more than 10% of the total peak area. Its peak area ratio was specified to be within 1.081–2.009.

Evaluation of Similarity

The data obtained from the chromatograms was converted and the useful data was extracted and input into the Similarity Degree Calculation Software for Traditional Chinese Medicine Fingerprints analysis system. The differences of peaks were evaluated and the overall similarity was analyzed. The degree of similarity was 0.92–0.95 for

Chuanxiong collected from different locations in Pengzhou, whilst it was 0.72–0.81 for samples collected from different origins. Therefore the degree of similarity was high for Chuanxiong of the same origin, but there were good correlations and yet differences for samples originating from different locations.

2.7

Nüzhenzi
女贞子
Fructus ligustri lucidi

Introduction

Fructus ligustri lucidi (family Oleaceae) is the dried fruit of *Ligustrum lucidum* Ait.

It mainly originates in Zhejiang, Jiangsu, Fujian, Hunan, Sichuan and Guangxi Provinces. According to studies on resources, it is planted abundantly and produced in large quantities in Mianyang, Zigong, Chengdu, Daxian, Dazhou, Neijiang, Rongxian, Renshou, Anyue, Pengan, Yingshan, Xichong, Pingchang, Guangan, Kaijiang, Leshan, Jiajiang, Dayi, Huili, Xide, Leibo, Yibin and Luzhou of Sichuan Province.

Chemical Composition

The peel of *Fructus ligustri lucidi* contains about 14% oleanolic acid, acetyloleanolic acid, malol, mannitol and dextrose. The fruit of *Fructus ligustri lucidi* contains nuzhenide. The seed of *Fructus ligustri lucidi* contains 16.9% fatty oils, of which 19.5% are palmitic acid and stearic acid, and 80.5% are oleic acid and linoleic acid.

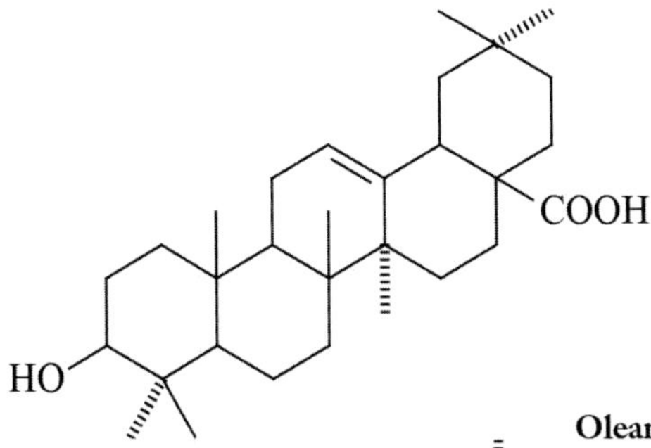

Oleanolic acid

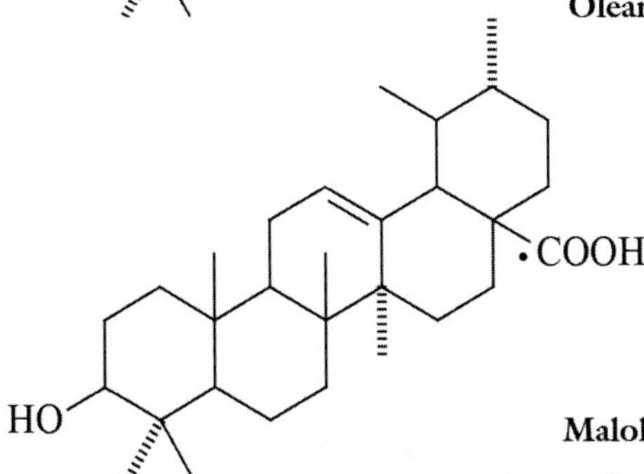

Malol

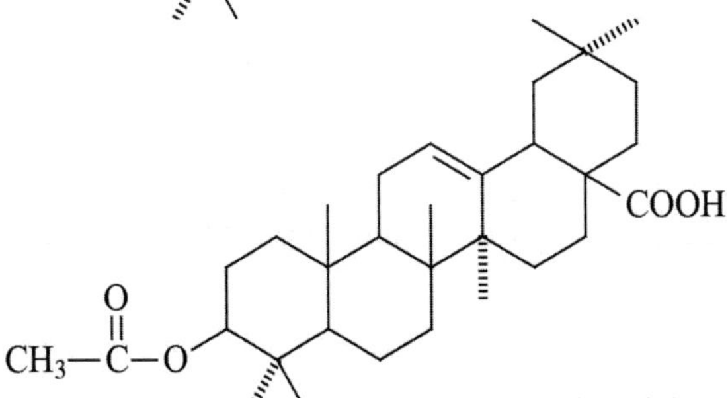

Acetyloleanolic acid

Experiment Materials

Samples

Fructus ligustri lucidi samples were collected from Daxian in Sichuan in the second half of December 2002. There were 12 samples identified as *Fructus ligustri lucidi.*

Reagents

Acetonitrile (chromatographic-grade, Tedia Company Inc., USA). Double-distilled deionized water (prepared by our lab). Ethanol (analytical-grade ethanol was purchased from Shanghai Chemical Reagents Pte. Ltd.; chromatographic-grade ethanol was purchased from Tedia Company Inc.).

Preparation of Samples

1 g of crude powder sample was accurately weighed. 50 ml of ethanol was added and refluxed for 30 min. After cooling, it was filtered and the residue was added with ethanol and refluxed two more times. The filtrates were combined and evaporated. The residue was dissolved and topped up to 10 ml with warm absolute ethanol in a 10 ml volumetric flask. After shaking, the solution was filtered through a 0.45 μm Millipore filter. The filtrate was then ready to be used as a test sample.

HPLC Conditions

Instrument: Agilent 1100 high-performance liquid chromatography and diode array detector
Mobile phase: Water and acetonitrile with gradient elution
Flow rate: 1.0 ml/min
Column: Agilent Zorbax 80A Extend-C_{18} analytical column (250 mm × 4.6 mm, 5 μm)
Column temperature: 25°C
Detecting wavelength: 220 nm
Reference wavelength: 360 nm
Run time: 80 min
Injection volume: 20 μl

Methods

The sample solution was injected into the liquid chromatograph. The chromatogram was obtained under selected HPLC conditions. The experimental method was validated.

Mobile Phase Selection

A comparison study showed that gradient elution using water (A) and acetonitrile (B) could successfully separate the peaks and could produce the maximum number of peaks from the samples. The duration of the analysis was 80 min. The 2 h chromatogram showed no characteristic peaks after 80 min.

A is water, B is acetonitrile.

	A (%)	B (%)
0–25 min	80–98	2–20
25–45 min	45–80	20–55
45–75 min	10–45	55–90
75–80 min	10–10	90–90

Method Validation

1. Stability: Samples from the same batch were stored at room temperature and injected at regular intervals. The degrees of similarity of the injections calculated by the Similarity Degree Calculation Software for Traditional Chinese Medicine Fingerprints were all above 0.92, which indicated that samples were stable within the 36 testing hours.

2. Precision: Five consecutive injections of the samples from the same batch were conducted to determine the precision. The degrees of similarity of the injections calculated by the Similarity Degree Calculation Software for Traditional Chinese Medicine Fingerprints were all above 0.95, which indicated that the instrument is of acceptable precision.

3. Reproducibility: Five samples from the same batch were prepared and analyzed according to the method described above. The degrees of similarity of the injections calculated by the Similarity Degree Calculation Software for Traditional Chinese Medicine Fingerprints were all above 0.91, indicating good reproducibility of the results.

Sample Analysis

The HPLC fingerprints of the test samples from 10 batches were analyzed under the conditions described above.

Establishment of Fingerprints

The number of peaks, peak area and relative retention time of 10 batches of samples were analyzed and compared to obtain an optimized fingerprint (Figs. 7.1 and 7.2).

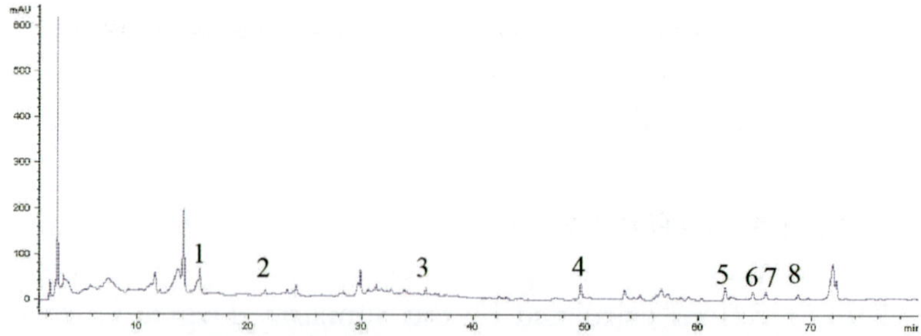

Fig. 7.1. HPLC fingerprint of Nüzhenzi (*Fructus ligustri lucidi*).

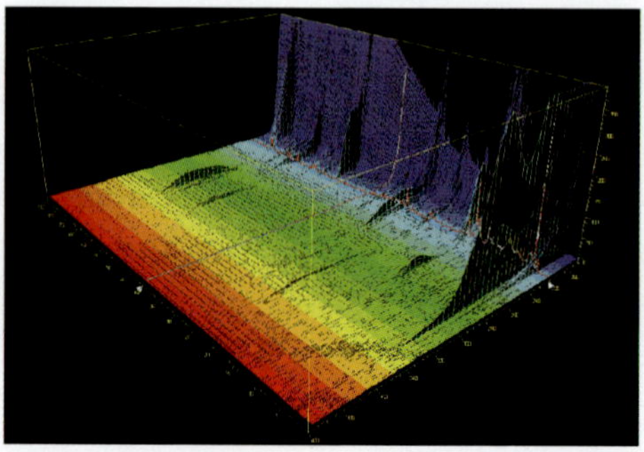

Fig. 7.2. 3D-HPLC fingerprint of Nüzhenzi (*Fructus ligustri lucidi*).

Results and Discussion

Over 20 peaks were separated by HPLC. After comparing all the tested and documented fingerprints, eight common peaks were selected as the characteristic peaks. The retention time (min) and peak area values of the eight common peaks are listed below.

Peak number	1	2	3	4	5	6	7	8
Retention time	15.612	21.38	35.685	49.555	62.306	64.772	65.927	68.768
Peak area	1770.6	317.1	309.3	690.8	566.3	280.2	325.7	266.1

Common Peak Standardization

Common peaks were assigned according to the results of 10 batches of samples. There were eight common peaks in the HPLC fingerprints of *Fructus ligustri lucidi*. The common peak with a retention time of about 49.55 min was subsequently used as the reference peak. Compared to the reference peak (peak 4), the relative retention times of the other seven common peaks were 0.315, 0.431, 0.720, 1.257, 1.307, 1.330 and 1.388.

Ratio of Common Peak Area

No common peak area was more than 10% of the total peak area.

2.8

Tianma
天麻
Rhizoma gastrodiae

Introduction

Rhizoma gastrodiae (family Orchidaceae) is the dried stem of *Gastrodia elata* Bl.

It mainly originates from Sichuan, Yunnan, Hubei, Shaanxi and Guizhou Provinces. It was originally wild, but is now cultivated everywhere. Tianma produced in Guizhou is the most famous one for medication. Wild Tianma from Zhaotong of Yunnan Province is famous for its good quality. Most of the commercially cultivated Tianma is from Shaanxi Province.

According to studies on resources, it is planted abundantly in Jinzhai, Yuexi and Huoshan of Anhui Province.

Chemical Composition

It mainly contains three types of ingredients:

1. Phenols and glycosides: It contains vanillyl alcohol, gastrodin and aglucon, such as *p*-hydroxybenzyl alcohol, *p*-hydroxybenzaldehyde, 3,4-dihydroxybenzaldehyde, 4,4′-dihydroxydiphenyl methane, *p*-hydroxybenzyl ethyl ether, 4,4′-dihydroxydiphenyl ether, 4-ethyloxytolyl-4′-hydroxybenzyl ether, tris[4-β-D-glucopyranosyloxybenzyl]-citrate, parishin, 4-hydroxybenzyl

methyl ether, 4-[4′-hydroxybenzyloxyl]-benzyl methyl ether, bis[4-hydroxybenzyl]-ether-mono-β-D-glucopyranoside and gastrodioside.

2. Sterols and organic acid: It contains β-sitosterol, daucosterol, citric acid and succinic acid.

3. Polysaccharides: It contains three kinds of polysaccharides, which are GE-I (relative molecular weight 3,000), GE-II (relative molecular weight 3,100) and GE-III (relative molecular weight 2,210). All of them are white powders. According to element analysis, nitrogen is absent in these polysaccharides, and they are mainly made up of glucose. Some of the chemical structures are listed below.

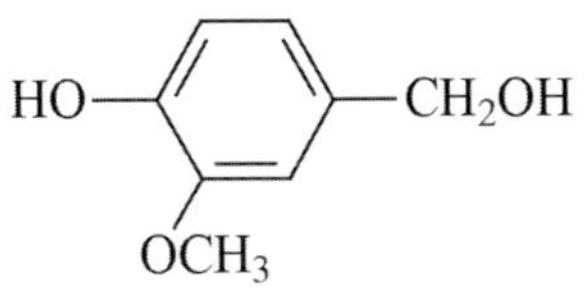

Vanillyl alcohol

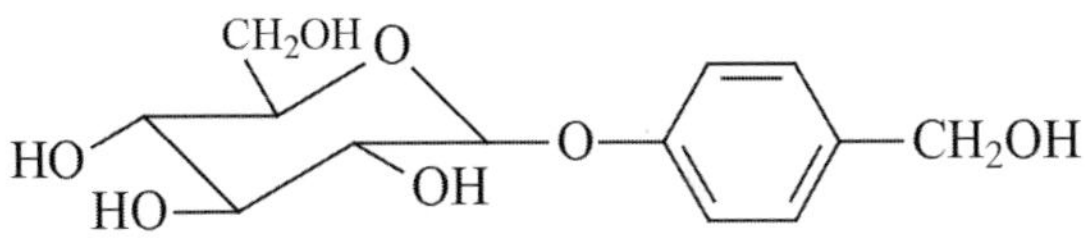

Gastrodin

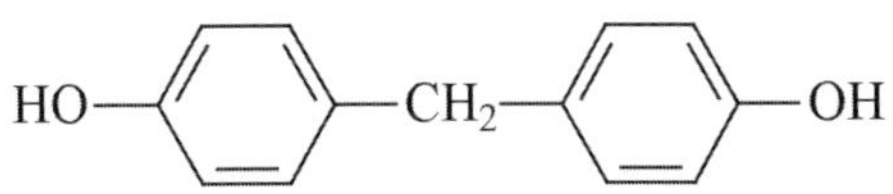

4,4'-Dihydroxydiphenyl methane

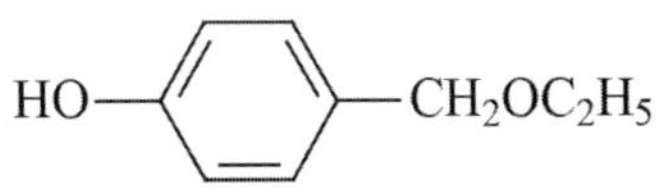

p-Hydroxybenzyl ethyl ether

4,4'-Dihydroxydiphenyl ether

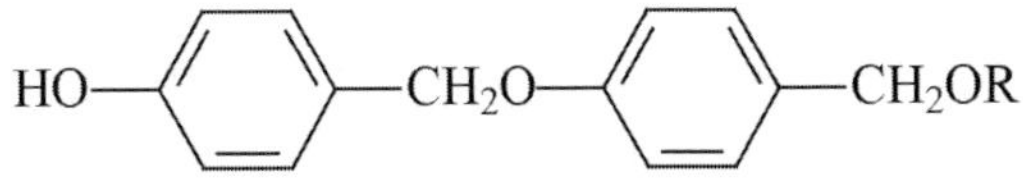

4-Ethyloxytolyl-4′-hydroxybenzyl ether R=C_2H_5
4-[4′-Hydroxybenzyloxyl]-benzyl methyl ether R=CH_3

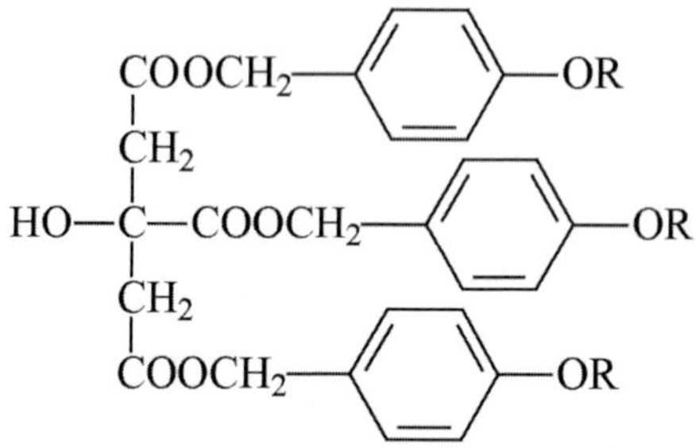

Tris[4-β-D-glucopyranosyloxybenzyl]-citrate R=glucose

Experiment Materials

Samples

Rhizoma gastrodiae samples were collected from three-year-old cultivated plants in Jinzhai of Anhui Province in December 2002 (Winter *Rhizoma gastrodiae*). There were 13 samples identified as *Rhizoma gastrodiae*.

Reference

Gastrodin used for quantitative analysis was provided by the National Institute for the Control of Pharmaceutical and Biological Products (China).

Reagents

Acetonitrile (chromatographic-grade, Tedia Company Inc., USA). Methanol (analytical-grade methanol was purchased from Shanghai Chemical Reagents Pte. Ltd.; chromatographic-grade methanol was purchased from Tedia Company Inc.). Phosphoric acid (HPLC-grade, Tedia Company Inc., USA). Double-distilled deionized water (prepared by our lab).

Preparation of Samples

An appropriate amount of crude powder was dried at 60°C for 6 h, then ground to pass through an 80 mesh sieve. 0.5 g of the ground powder sample was accurately weighed. 10 ml of 60% methanol was added and sonicated for 60 min with an operating frequency of 40 kHz and power of 500 W. After 2 h, the supernatant was collected and filtered through a 0.45 μm Millipore filter. The filtrate was then ready for analysis.

Preparation of Reference

An appropriate amount of gastrodin was accurately weighed and dissolved in methanol to obtain a concentration of 0.102 mg/ml. It was then ready to be used as a reference.

HPLC Conditions

Instrument: Agilent 1100 high-performance liquid chromatography and variable wave detector
Mobile phase: Water and acetonitrile with gradient elution
Flow rate: 1.0 ml/min
Column: Lichrospher C_{18} analytical column (250 mm × 4.6 mm, 5 μm)
Column temperature: 25°C
Detecting wavelength: 226 nm
Reference wavelength: 360 nm
Run time: 65 min
Injection volume: 20 μl

Methods

The sample solution was injected into the liquid chromatograph. The chromatogram was obtained under selected HPLC conditions. The experimental method was validated.

Mobile Phase Selection

A comparison study showed that gradient elution using 0.1% H_3PO_4 in water (A) and 0.1% H_3PO_4 in methanol (B) could successfully separate the chromatographic peaks with the maximum number of peaks. The duration of the analysis was 65 min. The 1.5 h chromatogram showed no characteristic peaks after 65 min.

A is 0.1% H_3PO_4 in water, and B is 0.1% H_3PO_4 in methanol.

	A (%)	B (%)
0–15 min	93–97	3–7
16–65 min	0–93	7–100

Method Validation

1. Stability: Samples from the same batch were stored at room temperature and injected at regular intervals. The degrees of similarity of the injections calculated by the Similarity Degree Calculation Software for Traditional Chinese Medicine Fingerprints were all above 0.91, which indicated that the samples were stable within the 36 testing hours.
2. Precision: Five consecutive injections of the samples from the same batch were made to determine the precision. The degrees of similarity of the injections calculated by the Similarity Degree Calculation Software for Traditional Chinese Medicine Fingerprints were all above 0.94, which indicated that the instrument is of acceptable precision.
3. Reproducibility: Five samples from the same batch were prepared and analyzed according to the method described above. The degrees of similarity of the injections calculated by the Similarity Degree Calculation Software for Traditional Chinese Medicine Fingerprints were all above 0.93, indicating good reproducibility of the results.

Sample Analysis

HPLC analysis was conducted for 10 batches of samples under the conditions described above.

Establishment of Fingerprints

The number of peaks, peak area and relative retention time of 10 batches of samples were analyzed and compared to obtain an optimized fingerprint (Fig. 8.1).

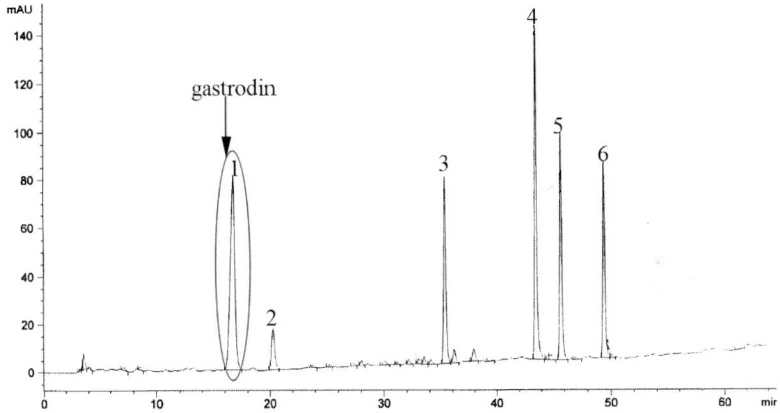

Fig. 8.1. HPLC fingerprint of Tianma (*Rhizoma gastrodiae*).

Results and Discussion

Common Peak Standardization

Common peaks were assigned according to the results of the 10 batches of samples. There were six common peaks in the HPLC fingerprints of *Rhizoma gastrodiae*. Compared to the reference peak (peak 1), the relative retention times of the other five common peaks were 1.211, 2.111, 2.598, 2.731 and 2.954.

Ratio of Common Peak Area

Aside from the reference peak, the peak areas for peaks 3, 4, 5, 6 exceeded 10% of the total peak area. The ratio of peaks 3, 4, 5, 6 to the reference peak was specified to be within 0.335–0.623, 0.547–1.015, 0.365–0.679 and 0.307–0.569 respectively.

2.9

Wuweizi

五味子

Fructus schisandrae chinensis

Introduction

Fructus schisandrae chinensis (family Magnoliaceae) is the dried fruit of *Schisandra chinensis* (Turcz.) Baill. It is commonly called Northern Wuweizi. It mainly originates from Liaoning, Heilongjiang and Jilin. According to studies on resources, it grows in abundance in the wild in Wuchang, Shangzhi, Acheng, Shuangcheng, Fangzheng, Yilan, Jiamusi, Yichun, Huanan, Boli, Qitaihe, Linkou, Nanning, Dongning, Baoqing, Qingan, Tieli, Hulin and Bei'an, in Heilongjiang Province.

Chemical Composition

It mainly contains volatile oils and lignanoids. The lignanoids include schizandrin, schizandrin A, schizandrin B, γ-schizandrin, pseudo schizandrin, schizandrols A and B, schizandrin C, wuweizi ethers A and B, and schizanhenol. Their structures are listed below.

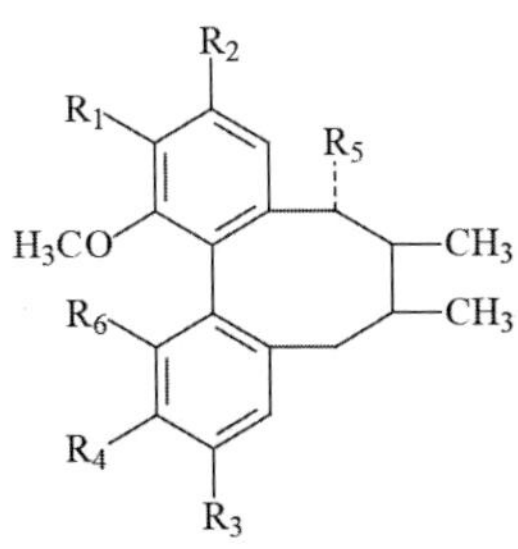

Schizandrin $R_1=R_2=R_3=R_4=R_6=OCH_3$, $R_5=OH$
Schizandrin A $R_1=R_2=R_3=R_4=R_6=OCH_3$, $R_5=H$
Schizandrin B $R_1=R_2=R_6=OCH_3$, $R_5=OH$, R_3—R_4=O-CH_2-O
Pseudo schizandrin $R_1=R_2=R_3=OCH_3$, R_4—R_6=O-CH_2-O
Schizandrol A $R_1=R_3=R_4=R_6=OCH_3$, $R_2=R_5=OH$
Schizanhenol $R_1=R_2=R_3=R_4=OCH_3$, $R_2=H$, $R_5=OH$
Schizandrin C R_1—R_2=O-CH_2-O, R_3—R_4= O-CH_2-O, $R_5=H$, $R_6=OCH_3$
Schizandrol B R_1—R_2=O-CH_2-O, $R_3=R_4=R_6=OCH_3$, $R_5=OH$

Experiment Materials

Samples

Fructus schisandrae chinensis samples were purplish-red fruit collected from three-year old wild plants in Qingan, Tieli, etc, of Heilongjiang Province in the second half of October 2001. There were 12 samples identified as *Fructus schisandrae chinensis.*

References

Schizandrol A, schizandrins A and B used for quantitative analysis were provided by the National Institute for the Control of Pharmaceutical and Biological Products (China).

Reagents

Acetonitrile (chromatographic-grade, Tedia Company Inc., USA). Methanol (analytical-grade methanol was purchased from Shanghai Chemical Reagents Pte. Ltd.; chromatographic-grade methanol was purchased from Tedia Company Inc.). Double-distilled deionized water (prepared by our lab).

Preparation of Samples

1 g of crude powder (passed through a 60 mesh sieve) was extracted in an appropriate amount of methanol and sonicated for 20 min twice with an operating frequency of 40 kHz and power of 500 W. After filtering, the supernatant was combined and made up to 25 ml with methanol. The solution was then ready for analysis.

Preparation of Reference

0.042 mg of schizandrol A, 0.048 mg of schizandrin A and 0.045 mg of schizandrin B were accurately weighed and dissolved in 1 ml of methanol. It was then ready to be used as a reference.

HPLC Conditions

Instrument: Agilent 1100 high-performance liquid chromatography and diode array detector
Mobile phase: Water and acetonitrile with gradient elution
Flow rate: 1.0 ml/min
Column: Kromasil® KR100-5C_{18} analytical column (250 mm × 4.6 mm, 5 μm)
Column temperature: 30°C
Detecting wavelength: 240 nm
Reference wavelength: 380 nm
Run time: 70 min
Injection volume: 10 μl sample, 15 μl reference

Methods

Mobile Phase Selection

A comparison study showed that gradient elution using water (A) and acetonitrile (B) could successfully separate into the maximum number of peaks. The duration of the analysis was 70 min. The 2 h chromatogram showed no characteristic peaks after 70 min.

A is water and B is acetonitrile.

	A (%)	B (%)
0–15 min	98–45	2–55
15–30 min	45–40	55–60
30–50 min	70–60	60–70
50–60 min	70–0	30–100
60–70 min	0–0	100–100

Method Validation

1. Stability: Samples from the same batch were stored at room temperature and injected at regular intervals. The degrees of similarity of the injections calculated by the Similarity Degree Calculation

Software for Traditional Chinese Medicine Fingerprints were all above 0.9, which indicated that samples were stable within the 36 testing hours.

2. Precision: Five consecutive injections of the samples from the same batch were conducted to determine the precision. The degrees of similarity of the injections calculated by the Similarity Degree Calculation Software for Traditional Chinese Medicine Fingerprints were all above 0.95, which indicated that the instrument is of acceptable precision.
3. Reproducibility: Five replicates from the same batch were prepared and analyzed according to the method described above. The degrees of similarity of the injections calculated by the Similarity Degree Calculation Software for Traditional Chinese Medicine Fingerprints were all above 0.95, indicating good reproducibility of the results.

Sample Analysis

HPLC analysis was conducted for 10 batches of samples under the conditions described above.

Establishment of Fingerprints

The number of peaks, peak area and relative retention time of 10 batches of samples were analyzed and compared to obtain an optimized fingerprint (Fig. 9.1).

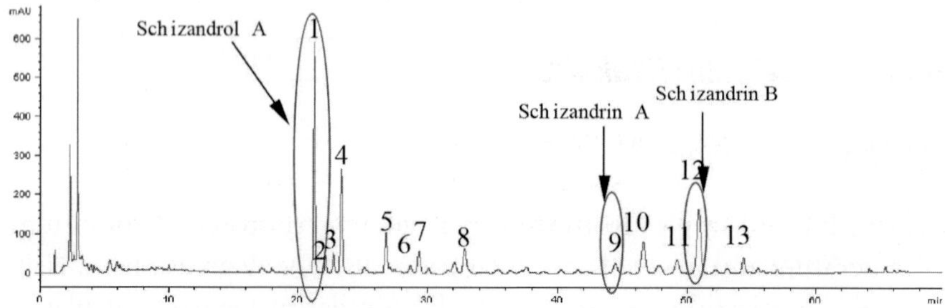

Fig. 9.1. HPLC fingerprint of Wuweizi (*Fructus schisandrae chinensis*).

Results and Discussion

About 40 peaks were separated under the above HPLC conditions. Thirteen common peaks were selected to characterize the fingerprint of *Fructus schisandrae chinensis* by comparing all peaks obtained from the samples and the reference standards. The retention time (min) and peak area values are listed below.

Peak number	1	2	3	4	5	6	7
Retention time	21.265	22.166	22.775	23.381	26.818	28.669	29.311
Peak area	5196.7	475.2	547.1	2852.6	1478.5	411.2	902.8

Peak number	8	9	10	11	12	13	
Retention time	32.825	44.397	46.546	49.172	50.842	54.344	
Peak area	1221.5	754.8	2180.1	1117.3	3892.2	837.6	

Common Peak Standardization

Common peaks were assigned according to the results of 10 batches of samples. There were 13 common peaks in the HPLC fingerprints of *Fructus schisandrae chinensis*. Compared to the reference peak of schizandrol A (peak 1), the relative retention times of the other 12 common peaks were 1.042, 1.071, 1.100, 1.261, 1.348, 1.378, 1.544, 2.088, 2.189, 2.312, 2.391 and 2.556.

Ratio of Common Peak Area

Aside from the reference peak for schizandrol A, the peak area for peak 12 (schizandrin B) was more than 10% of the total peak area. The ratio of peak 12 to the reference peak was specified to be within 0.524–0.974.

2.10

Taizishen
太子参
Radix pseudostellariae

Introduction

Radix pseudostellariae (family Caryophyllaceae) is the dried root of *Pseudostellaria heterophylla* (Miq.) Pax ex Pax et Hoffm. It is mainly produced in Jiangsu, Shandong and Anhui Provinces. According to studies on resources, it is cultivated in large quantities in Jurong, Lishui, Liyang, Yixing, Gaochun, Jiangning and Jiangpu of Jiangsu Province.

Chemical Composition

Radix pseudostellariae mainly contains saponins and volatile oils. Twenty-seven compounds, such as palmitic acid, linoleic acid, tripalmitin, β-sitosterol, Δ^7-stigmast-3-β-ol, daucosterol, Δ^7-stigmasterol-3-β-D-glucopyranoside, pseudostellaria cyclopeptides A, B, C, and pseudostellarinoside A were separated and identified after extracting volatile oils, trace elements and amino acids from the *Radix pseudostellariae*.

Experiment Materials

Samples

Radix pseudostellariae samples were collected from one-year-old cultivated plants in Yixing, Gaochun, Jurong and Jiangning of Jiangsu Province in October 2002. There were 14 samples identified as *Radix pseudostellariae*.

Reagents

Acetonitrile (chromatographic-grade, Tedia Company Inc., USA). Methanol (analytical-grade methanol was purchased from Shanghai Chemical Reagents Pte. Ltd.; chromatographic-grade methanol was purchased from Tedia Company Inc.). Ether and *n*-butanol (analytical-grade, Shanghai Chemical Reagent Pte. Ltd.). Double-distilled deionized water (prepared by our lab).

Preparation of Samples

2 g of crude powder was Soxhlet extracted with ether in a water bath at 40°C to defat for 2 h. The ether was evaporated off and the residue was Soxhlet extracted with methanol at 80°C in a water bath for 3–4 h until the acetic anhydride sulfuric acid test was negative. The methanol extract was dried under vacuum and 30 ml of distilled water was added to dissolve the residue. The mixture was then extracted with 60 ml of water-saturated *n*-butanol and the extract was washed with *n*-butanol-saturated water. The *n*-butanol was evaporated under vacuum and the residue was dissolved in methanol and topped up to 10 ml in a volumetric flask. The solution was filtered with a 0.45 nm Millipore filter. The solution was then ready for analysis.

HPLC Conditions

Instrument: Agilent 1100 high-performance liquid chromatography and diode array detector
Mobile phase: Water and acetonitrile with gradient elution
Flow rate: 1.0 ml/min
Column: Agilent Zorbax 80A Extend-C_{18} analytical column (250 mm × 4.6 mm, 5 μm)
Column temperature: 25°C
Detecting wavelength: 202 nm
Reference wavelength: 360 nm
Run time: 70 min
Injection volume: 20 μl

Methods

The sample was injected into the liquid chromatograph. The chromatogram was obtained under selected HPLC conditions. The experimental method was validated.

Mobile Phase Selection

A comparison study showed that gradient elution using water (A) and acetonitrile (B) could successfully separate the peaks and could produce the maximum number of peaks from the samples. The duration of the analysis was 70 min. The 2 h chromatogram showed no characteristic peaks after 70 min.

A is water and B is acetonitrile.

	A (%)	B (%)
0–10 min	90–98	2–10
10–30 min	40–90	10–60
30–65 min	10–40	60–90
65–70 min	10–10	90–90

Method Validation

1. Stability: Samples from the same batch were stored at room temperature and injected at regular intervals. The degrees of similarity of the injections calculated by the Similarity Degree Calculation Software for Traditional Chinese Medicine Fingerprints were all above 0.94, which indicated that samples were stable within the 36 testing hours.

2. Precision: Five consecutive injections of the samples from the same batch were conducted to determine the precision. The degrees of similarity of the injections calculated by the Similarity Degree Calculation Software for Traditional Chinese Medicine Fingerprints were all above 0.96, which indicated that the instrument is of acceptable precision.

3. Reproducibility: Five replicates from the same batch were prepared and analyzed according to the method described above. The degrees of similarity of the injections calculated by the Similarity Degree Calculation Software for Traditional Chinese Medicine Fingerprints were all above 0.91, indicating good reproducibility of the results.

Sample Analysis

HPLC analysis was conducted for 10 batches of samples under the conditions described above.

Establishment of Fingerprints

The number of peaks, peak area and relative retention time of 10 batches of samples were analyzed and compared to obtain an optimized fingerprint (Figs. 10.1, 10.2).

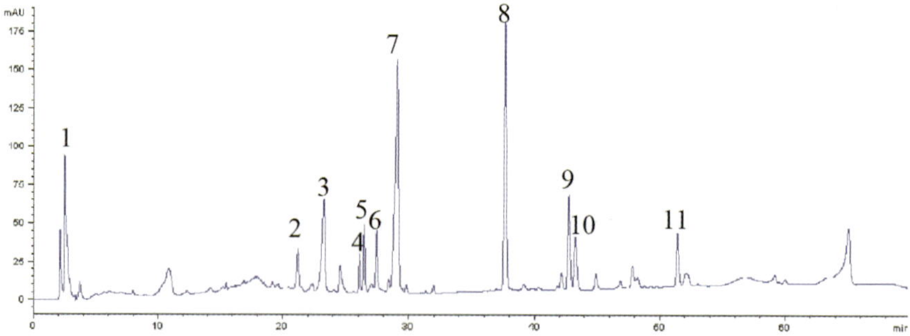

Fig. 10.1. HPLC fingerprint of Taizishen (*Radix pseudostellariae*).

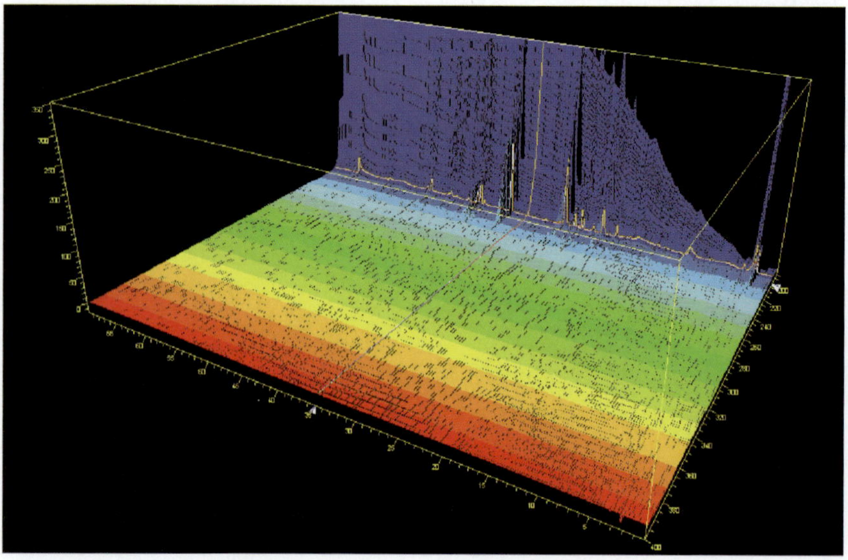

Fig. 10.2. 3-D HPLC fingerprint of Taizishen (*Radix pseudostellariae*).

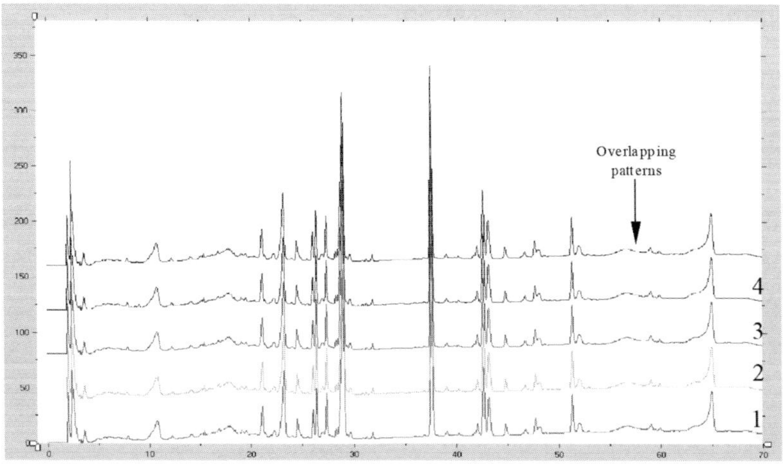

Fig. 10.3. Overlapping patterns of HPLC fingerprints of Taizishen (*Radix pseudostellariae*).

Results and Discussion

More than 20 peaks were separated under the HPLC conditions described above. After comparing all the tested and documented fingerprints, 11 common peaks were selected as the characteristic peaks of *Radix pseudostellariae*. Their retention time (min) and peak area values are listed below.

Peak number	1	2	3	4	5	6
Retention time	2.147	21.157	23.245	26.075	26.439	27.41
Peak area	1074.1	411.3	1197.3	204.6	366.8	395.6

Peak number	7	8	9	10	11
Retention time	29.008	37.6	42.689	43.208	51.373
Peak area	2887.5	2222.9	817.9	575.8	454

Common Peak Standardization

Common peaks were assigned according to the results of the 10 batches of samples. There were 11 common peaks in the HPLC fingerprints of *Radix pseudostellariae*. The common peak with a retention time of about 37.6 min was subsequently used as the reference peak. Compared to the reference peak (peak 8), the relative retention times of the other 10 common peaks were 0.064, 0.563, 0.618, 0.693, 0.703, 0.729, 0.771, 1.135, 1.149 and 1.366.

Ratio of Common Peak Area

Aside from the reference peak, the peak area of peak 7 was more than 10% of the total peak area. The ratio of peak 7 to the reference peak was specified to be within 0.909–1.689.

Evaluation of Similarity

The data obtained from the chromatogram was analyzed using the Similarity Degree Calculation Software for Traditional Chinese Medicine Fingerprints. The peaks were selected and matched automatically with matching templates assigned. After that, the standard matching template was specified and the differences among the peaks and the overall similarity were evaluated. Figure 10.3 shows the overlapping patterns of HPLC chromatograms of *Radix pseudostellariae*. The samples from Yixing, Ganchun, Jurong and Jiangning showed high similarity, of 0.99, 0.98, 0.96 and 0.98 respectively (between 0.96 and 0.99), which indicated that samples from the above four locations were highly correlated and ascertained that the quality of these samples was almost consistent.

2.11

Niuxi
牛膝
Radix achyranthis bidentatae

Introduction

Radix achyranthis bidentatae is the dried root of *Achyranthes bidentata* Bl. (family Amaranthaceae). It mainly originates from Henan Province. It is usually called Huai Niuxi (怀牛膝) and most of the plants are cultivated species.

According to some studies and investigations, *Radix achyranthis bidentatae* is planted abundantly in some areas of Henan Province, such as Wuzhi, Wenxian, Mengxian, Bo'ai, Qinyang and Huixian, which contribute to 60% of the total yields in China. The quality of the cultivated species in those areas is also the best in China, ensuring steady sales not only in China, but also in other countries.

Chemical Composition

Radix achyranthis bidentatae contains triterpenoid saponin and its aglucon, oleanolic acid. It also contains insect metamorphosis hormones, such as ecdysterone, inokosterone, epiecdysterone and rubrosterone.

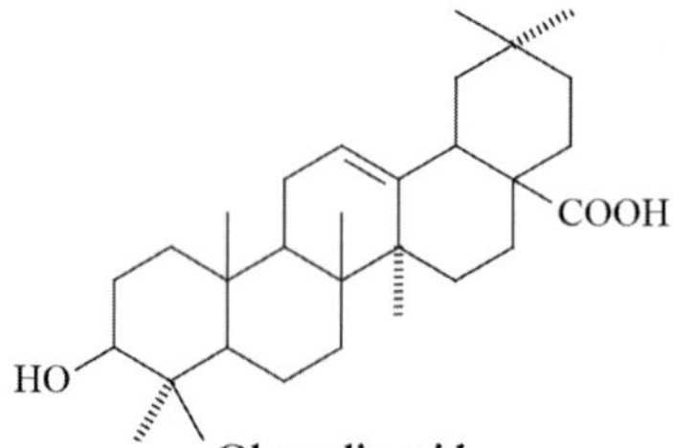

Oleanolic acid

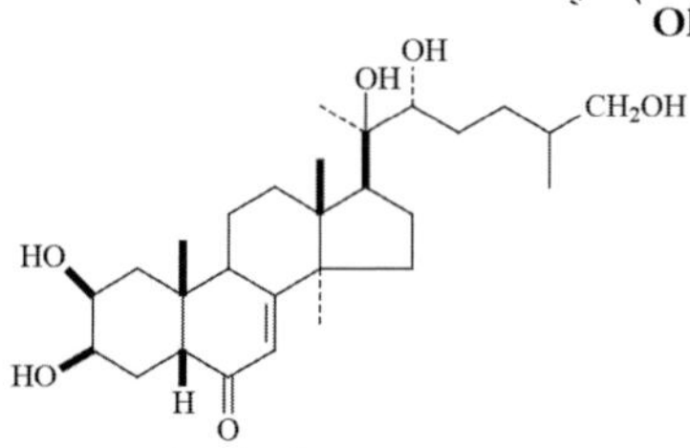

Inokosterone

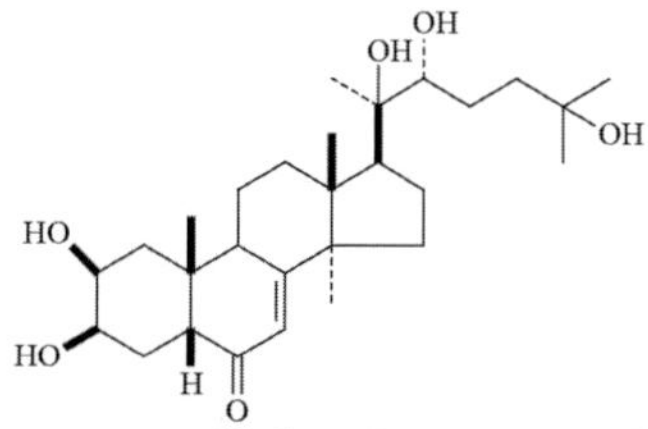

Ecdysterone

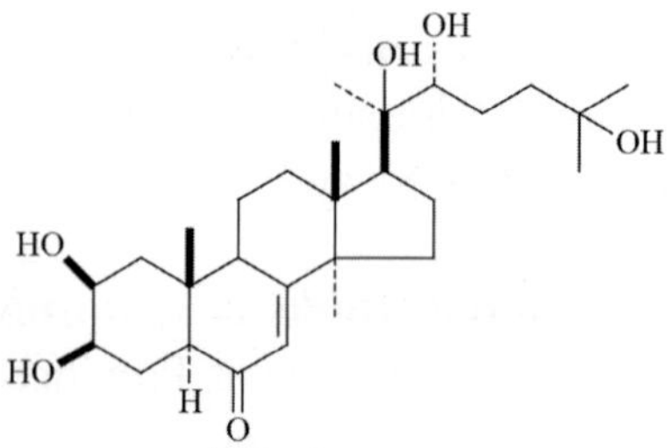

Epiecdysterone

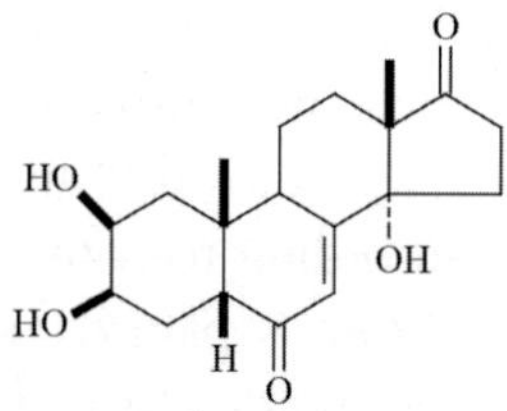

Rubrosterone

Experimental Materials

Samples

Radix achyranthis bidentatae samples were collected from one-year-old plants in Wuzhi (Henan Province) in winter 2002. There were 15 samples with variant identifications completed on all of them.

Reagents

Acetonitrile (chromatographic-grade, Tedia Company Inc., USA). Double-distilled deionized water (prepared by our lab). Methanol (analytical-grade methanol was purchased from Shanghai Chemical Reagents Pte. Ltd.; chromatographic-grade methanol was purchased from Tedia Company Inc.).

Preparation of Samples

2 g of crude powder of the sample was accurately weighed and placed in a round-bottom flask. 100 ml of methanol was added to the flask. After refluxing for 2 h, the sample solution was filtered. 100 ml of methanol was added to the residue and the sample was extracted for 1 h. After filtering, the two filtrates were combined. The filtrate was evaporated to dryness on an evaporating dish. The residue was dissolved and topped up to 10 ml with methanol in a volumetric flask. The mixture was filtered through a 0.45 μm Millipore filter. The filtrate was then ready to be used as a test sample.

HPLC Conditions

Instrument: Agilent 1100 high-performance liquid chromatography and diode array detector
Mobile phase: Water and acetonitrile with gradient elution
Flow rate: 1.0 ml/min
Column: Agilent ZORBAX 80A Extend-C_{18} analytical column (250 mm × 4.6 mm, 5 μm)
Column temperature: 25°C
Detecting wavelength: 250 nm
Reference wavelength: 360 nm
Run time: 65 min
Injection volume: 20 μl

Methods

The sample was injected into the liquid chromatograph. The chromatogram was obtained under selected HPLC conditions. The experimental method was validated.

Mobile Phase Selection

A comparison of experimental results showed that gradient elution using water and acetonitrile could successfully separate the peaks and produce the maximum number of peaks from the samples. The duration of the analysis was 65 min. The 1.5 h chromatogram of the samples showed no characteristic peaks after 65 min.

A is water, B is acetonitrile.

	A (%)	B (%)
0–10 min	90–98	2–10
10–45 min	60–90	10–40
40–60 min	20–60	40–80
60–65 min	20–20	80–80

Method Validation

1. Stability: Test samples from the same batch were stored at room temperature and injected at regular intervals for testing. The degrees of similarity of the injections calculated by the Similarity Degree Calculation Software for Traditional Chinese Medicine Fingerprints were all above 0.92, which indicated that samples were stable within the 36 testing hours.
2. Precision: Five consecutive injections of the samples from the same batch were made to determine the precision. The degrees of similarity of the injections calculated by the Similarity Degree Calculation Software for Traditional Chinese Medicine Fingerprints were all above 0.97, which indicated that the injections of the test samples were carried out with precision.
3. Reproducibility: Five samples from the same batch were prepared and tested according to the method described above. The degrees of similarity of the injections calculated by the Similarity Degree Calculation Software for Traditional Chinese Medicine Fingerprints were all above 0.94, indicating good reproducibility of the results.

Sample Analysis

The HPLC fingerprints of the test samples from 10 batches were analyzed under the conditions described above.

Establishment of Fingerprints

The number of peaks, peak area and relative retention time of 10 batches of samples were analyzed and compared to obtain an optimized fingerprint (Figs. 11.1, 11.2).

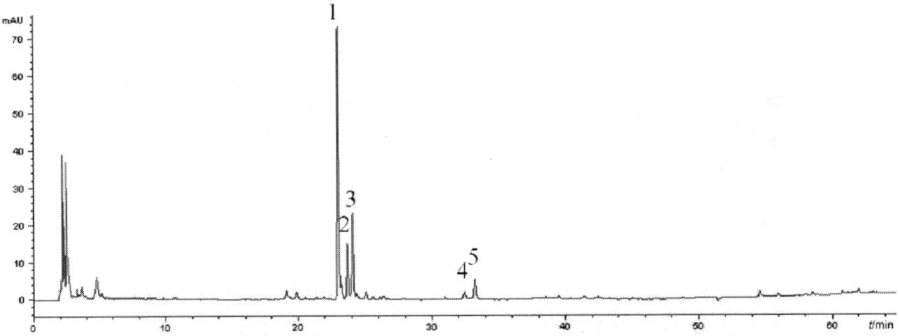

Fig. 11.1. HPLC fingerprint of Niuxi (*Radix achyranthis bidentatae*).

Results and Discussion

Over 20 peaks were separated using HPLC. After comparing all the tested and recorded fingerprints, five common peaks were selected as the stable characteristic peaks of the *Radix achyranthis bidentatae* fingerprint. The retention time (min) and peak area values of the five common peaks are listed below.

Peak number	1	2	3	4	5
Retention time	22.983	23.672	24.039	32.373	33.167
Peak area	583.4	116.5	186.4	19.4	53.2

Common Peak Standardization

Common peaks were assigned according to the results of 10 batches of samples. There were five common peaks in the HPLC fingerprints of *Radix achyranthis bidentatae*. The common peak with a retention time of about 22.98 min was subsequently used as the reference peak. Compared to the reference peak (peak 1), the relative retention times of the other four common peaks were 1.030, 1.046, 1.049 and 1.443.

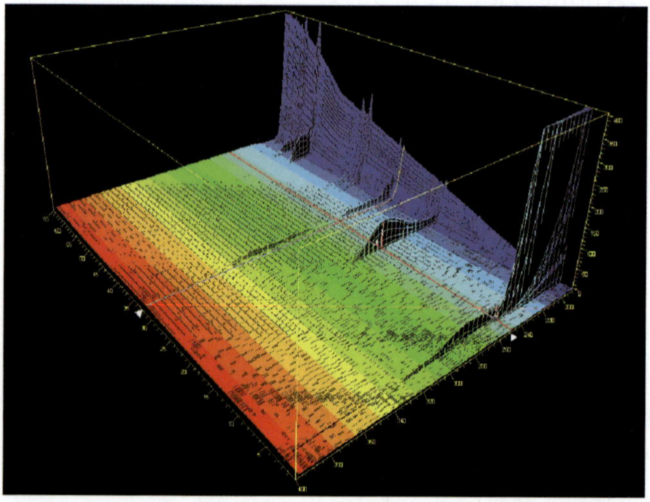

Fig. 11.2. 3-D HPLC fingerprint of Niuxi (*Radix achyranthis bidentatae*).

Ratio of Common Peak Area

Among the five common peaks, only the reference peak area exceeded 10% of the total peak area.

2.12

Danshen

丹参

Radix et Rhizoma Salviae miltiorrhizae

Introduction

Radix et Rhizoma Salviae miltiorrhizae is the dried root and rhizome of *Salvia miltiorrhiza* Bge. (family Labiatae). It mainly originates from Sichuan, Shanxi, Hebei, Jiangsu and Anhui Provinces.

According to studies on resources, the wild *Radix et Rhizoma Salviae miltiorrhizae* grows extremely well and in abundance in Shangluo, Luonan, Danfeng, Shangxian, Shanyang, Baihe and Pingli of Shaanxi Province. *Radix et Rhizoma Salviae miltiorrhizae* is planted abundantly in Shangluo Danshen Resource Base.

Chemical Composition

The main components of *Radix et Rhizoma Salviae miltiorrhizae* are liposoluble diterpene quinones and water-soluble phenolic acids. The liposoluble part contains tanshinones I, II_A, II_B, V, VI, cryptotanshinone, isotanshinones I, II_A, II_B, isocryoptotanshinone, hydroxytanshinone II_A, methyltanshinonate, salviol, dihydrotanshinone I, danshenxinkun A, danshenxinkun B and danshenxinkun C. The water-soluble phenolic acids mainly contain danshensu, salvianolic acids A, B, C, D, E, G and are shown on the next page protocatechualdehyde. Some of the chemical structures are shown on the next page.

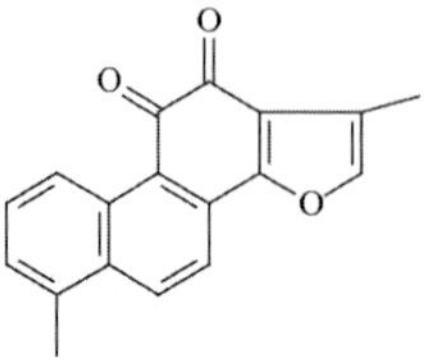

Tanshinone

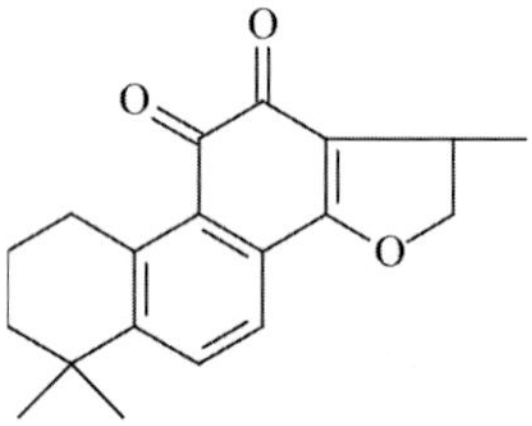

Cryptotanshinone

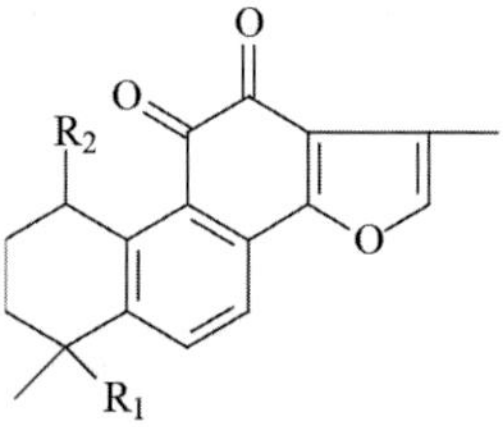

Tanshinone II_A $R_1=CH_3$, $R_2=H$
Tanshinone II_B $R_1=CH_2OH$, $R_2=H$
Methyltanshinonate
$R_1=COOCH_3$, $R_2=H$
Hydroxytanshinone II_A
$R_1=CH_3$, $R_2=OH$

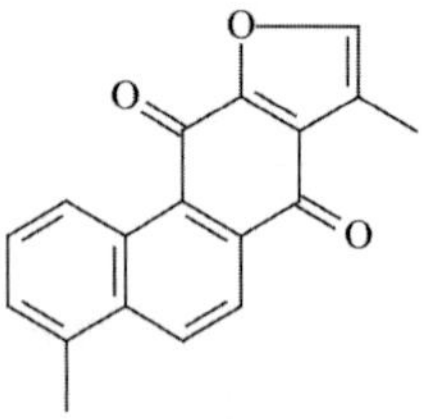

Isotanshinone I

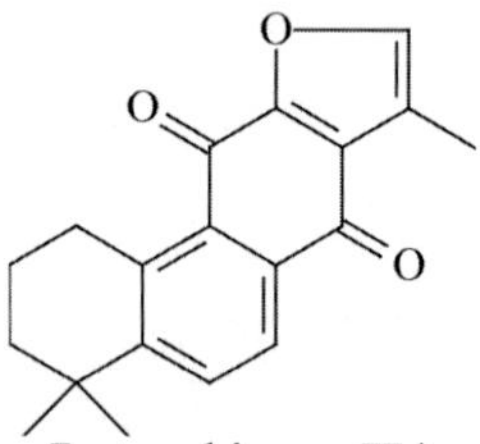

Isotanshinone IIA

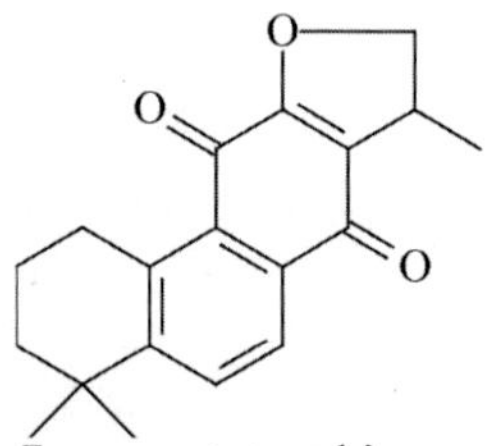

Isocryoptotanshinone

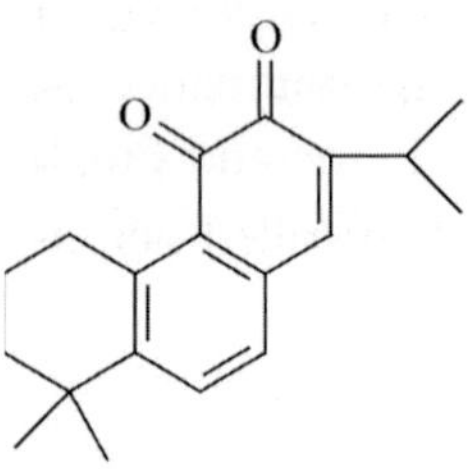

Dehydromiltirone

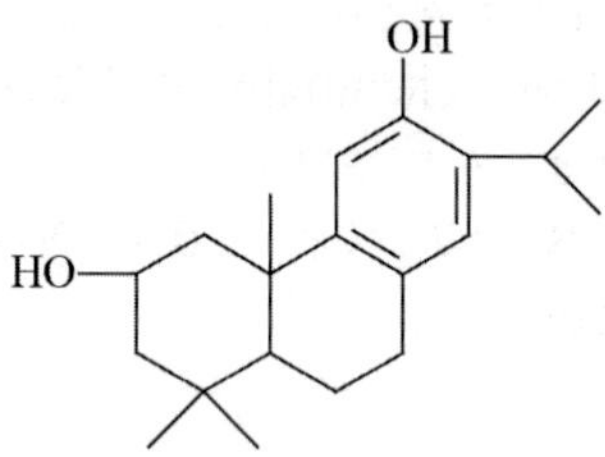

Salviol

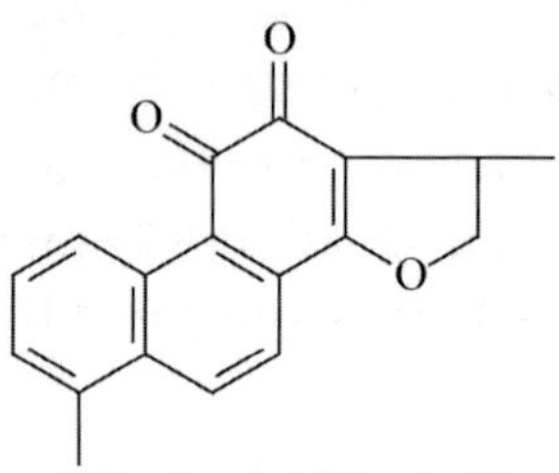

Dihydrotanshinone I

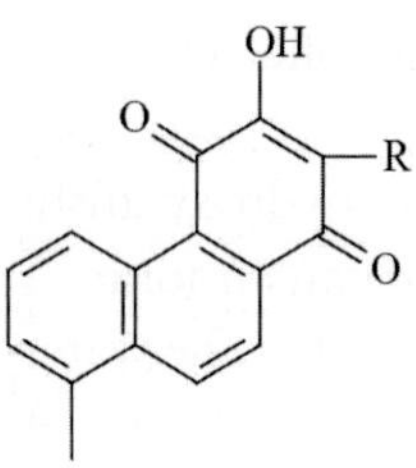

Danshenxinkun A R= —CH(CH_3)(CH_2OH)

Danshenxinkun B R= —CH(CH_3)(CH_3)

Danshenxinkun C **$R=CH_3$**

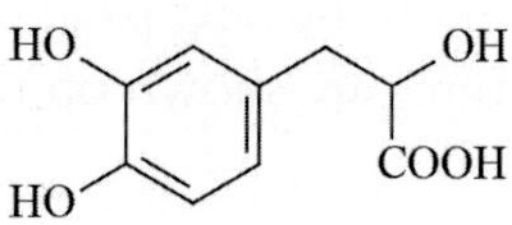

Danshensu

Experimental Materials

Samples

Radix et Rhizoma Salviae Miltiorrhizae samples were collected from the cultivated species in Tianlishi Shangluo Danshen Resource Base in Shangzhou of Shaanxi Province. They were collected in late November 2002. There were 12 samples with variant identifications completed on all of them.

References

Danshensu and tanshinone II_A used for quantitative analysis were all provided by the National Institute for the Control of Pharmaceutical and Biological Products (China).

Reagents

Methanol (analytical-grade methanol was purchased from Shanghai Chemical Reagents Pte. Ltd.; chromatographic-grade methanol was purchased from Tedia Company Inc.). Glacial acetic acid (analytical-grade, Nanjing Chemical Reagent Co. Ltd.). Double-distilled deionized water (prepared by our lab).

Preparation of Samples

1 g of powder sample was accurately weighed and placed in a Soxhlet extractor. 100 ml of water was added to the extractor. After extracting for 16 h, the volume of the residue was topped up to 100 ml. The sample solution was mixed well and filtered. The initial filtrate was discarded. The remaining filtrates were then ready to be used as test samples for analyzing water-soluble components.

0.5 g of powder sample was accurately weighed. 15 ml of methanol was added to the sample. Ultrasonic extraction was performed three times. The sample was extracted for 15 min every time using an operating frequency of 40 kHz and power of 500 W. The extracts were combined and topped up to 50 ml with methanol. The sample solution was mixed well and filtered. The initial filtrate was discarded. The remaining filtrates were then ready to be used as test samples for analyzing liposoluble components.

HPLC Conditions

Instrument: Agilent 1100 high-performance liquid chromatography and variable wave detector
Column: Hypersil ODS (4.6 mm × 250 mm, 5 μm)
Mobile phase: Methanol and 0.5% glacial acetic acid with gradient elution (for water-soluble components); methanol and water with gradient elution (for liposoluble components)
Flow rate: 0.6 ml/min (water-soluble components); 1 ml/min (liposoluble components)
Column temperature: 30°C
Detecting wavelength: 279 nm (water-soluble components); 270 nm (liposoluble components)
Run time: 60 min
Injection volume: 10 μl sample, 10 μl sodium danshensu reference, 20 μl tanshinone II_A reference

Methods

The sample solution was injected into the liquid chromatograph. The chromatogram was obtained under selected HPLC conditions. The experimental method was validated.

Mobile Phase Selection

A comparison of experimental results showed that when methanol and 0.5% glacial acetic acid were used as the mobile phase, gradient elution could successfully separate the peaks and produce the maximum number of peaks from the water-soluble samples. When methanol and water were used as the mobile phase, gradient elution could successfully separate the peaks and produce the maximum number of peaks from the liposoluble samples. The duration of the analysis was 60 min. The 1.5 h chromatogram of the samples showed no characteristic peaks after 60 min.

Water-soluble components: A is methanol, B is 0.5% glacial acetic acid.

	A (%)	B (%)
0–10 min	5–25	75–95
10–45 min	25–70	30–75
45–60 min	70–70	30–30

Liposoluble components: A is methanol, B is water.

	A (%)	B (%)
0–10 min	0–5	95–100
10–12 min	25–75	25–75
12–60 min	25–25	75–75

Method Validation

1. Stability: Water-soluble test samples from the same batch were tested at 0 h, 2 h, 4 h, 8 h, 24 h and 36 h. The RSD of the relative retention times of the main peaks was 2.62% and the RSD of the relative peak areas of the main common peaks was 2.45%, which meets the requirements. Liposoluble test samples from the same batch were tested at 0 h, 2 h, 4 h, 8 h and 16 h. The RSDs of all the main peaks' relative retention times and relative peak areas were less than 3%, which meets the requirement.

2. Precision: Five consecutive injections of the water-soluble samples from the same batch were conducted to determine the precision. The RSD of the relative peak areas of the main common peaks was 1.93%, which meets the requirement. Five consecutive injections of the liposoluble samples from the same batch were also conducted to determine the precision. All RSDs of the relative peak

areas of main common peaks were less than 2%, which meets the requirement.

3. Reproducibility: Five water-soluble and five liposoluble samples from the same batch were prepared according to the method described above. Parallel tests were performed on these samples. The RSDs of the relative retention times of all the main peaks were less than 2.5%, which meets the requirement.

Sample Analysis

The HPLC fingerprints of the samples from 10 batches were analyzed under the conditions described above.

Establishment of Fingerprints

The number of peaks, peak area and relative retention time of 10 batches of water-soluble and liposoluble test samples were analyzed and compared to obtain an optimized fingerprint (Figs. 12.1, 12.2).

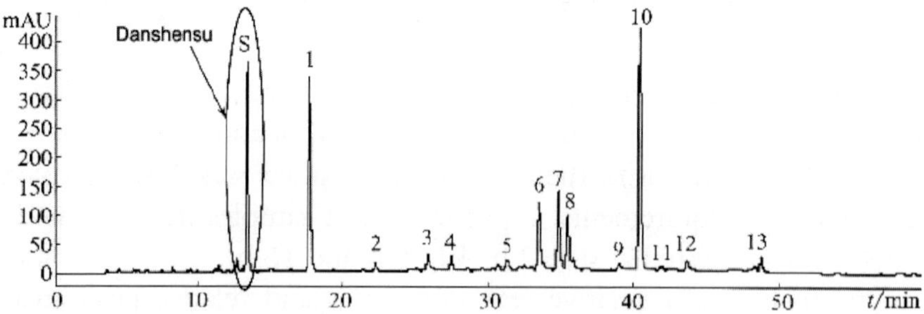

Fig. 12.1. HPLC fingerprint of water-soluble components of Danshen (*Radix et Rhizoma Salviae miltiorrhizae*).

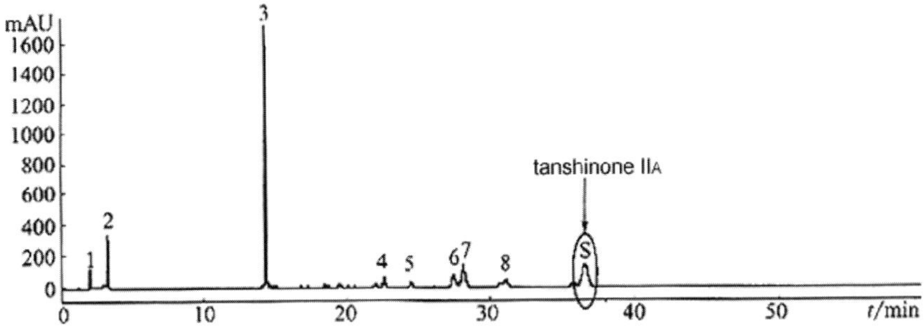

Fig. 12.2. HPLC fingerprint of liposoluble components of Danshen (*Radix et Rhizoma Salviae miltiorrhizae*).

Results and Discussion

Common Peak Standardization

Common peaks were assigned according to the results of 10 batches of samples. There were 14 common peaks in the HPLC fingerprints of the water-soluble components of *Radix et Rhizoma Salviae miltiorrhizae*. Compared to the danshensu reference peak (peak S), the relative retention times of the other 13 common peaks were 1.33, 1.68, 1.96, 2.08, 2.30, 2.54, 2.64, 2.69, 2.96, 3.07, 3.17, 3.31 and 3.70.

There were nine common peaks in the HPLC fingerprints of the liposoluble components of *Radix et Rhizoma Salviae miltiorrhizae*. Compared to the tanshinone II_A reference peak (peak S), the relative retention times of the other eight common peaks were 0.06, 0.09, 0.39, 0.61, 0.67, 0.75, 0.77 and 0.85.

Characteristics of Common Peaks

The HPLC fingerprints of the water-soluble components showed that when the column efficiency was high, peak 5 and peak 13 were further split into 2–3 peaks; peak 6 and peak 8 each had a shoulder peak or an incompletely separated small peak.

HPLC fingerprints of the liposoluble components showed that peak 8 has double peaks.

Ratio of Common Peak Areas

HPLC fingerprints of the water-soluble components of *Radix et Rhizoma Salviae miltiorrhizae* showed that the peak area of common peak 1 was more than 10% of the total peak area. Its ratio of relative peak areas was specified as 0.9–1.5.

HPLC fingerprints of the liposoluble components of *Radix et Rhizoma Salviae miltiorrhizae* showed that the peak areas of common peaks 2 and 3 were both more than 10% of the total peak area. Their ratios of relative peak areas were specified as 0.82–1.38 and 1.75–2.92, respectively.

2.13

Baimaogen
白茅根
Rhizoma imperatae

Introduction

Rhizoma imperatae is the dried rhizome of *Imperata cylindrica* Beauv. var. *major* (Nees) C.E. Hubb. (family Gramineae). It is planted all over China.

Chemical Composition

The rhizome of *Rhizoma imperatae* contains triterpenoids, such as 0.1% arundoin, 0.001% cylindrin, fernenol, arborane, isoarborinol, simiarenol, arborinol, arboriy-nolmethylether, arborinone, friedelin and 4,7-dimethoxy-5-methyl-coumarin. It also contains soluble calcium, 18.89% of sugars (mainly glucose, sucrose, a small amount of fructose and xylose), simple acids (such as citric acid and malic acid) and their potassium salts. It contains carotenoids, chlorophyll and anemonin.

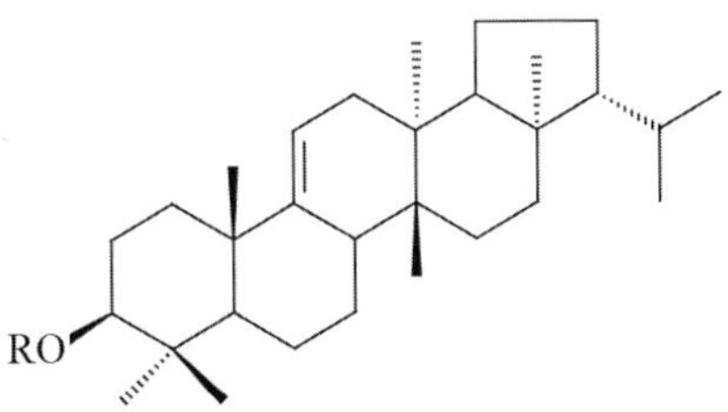

Arundoin $R=CH_3$
Fernenol $R=H$

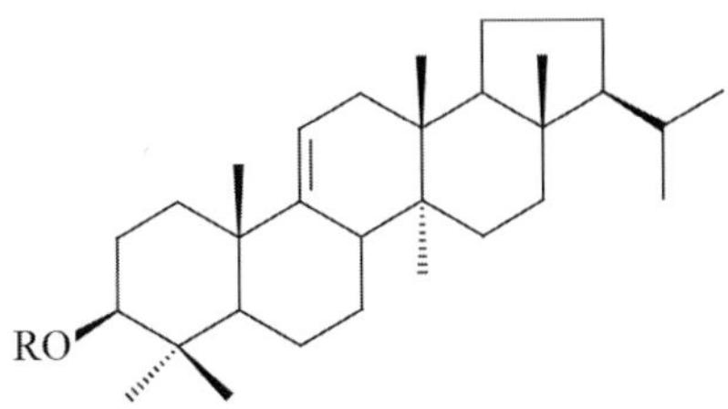

Cylindrin $R=CH_3$
Isoarborinol $R=H$

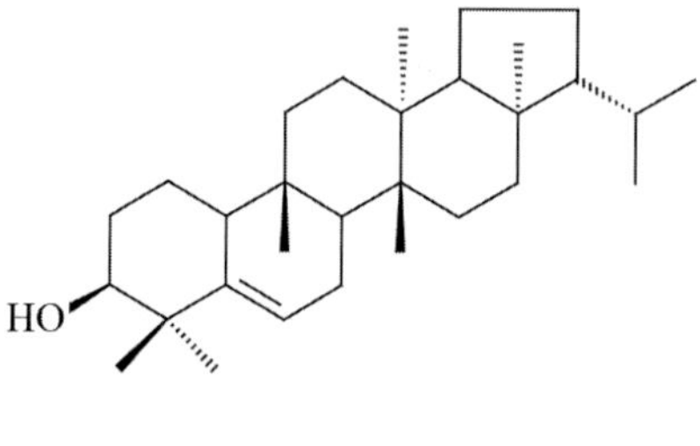

Simiarenol

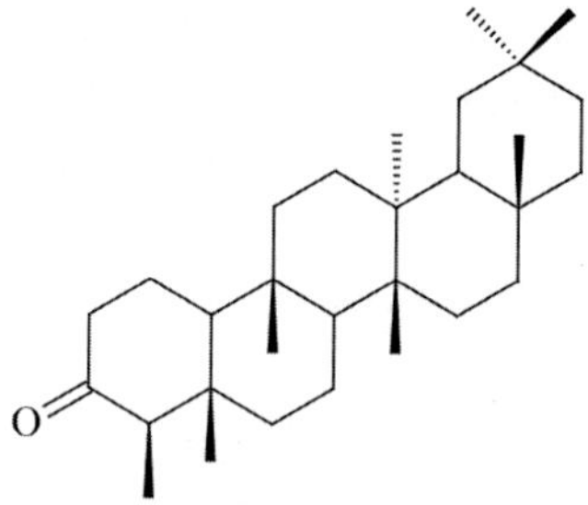

Friedelin

Experimental Materials

Samples

Rhizoma imperatae samples were collected from 10 places: Enshi and Huangshi of Hubei Province, Liuyang of Hunan Province, Yuncheng and Yuci of Shanxi Province, Baoding of Hebei Province, Nanyang and Zhengzhou of Henan Province, Hulunbeier Meng of Inner Mongolia, and Lanzhou of Gansu Province. The samples all underwent variant identification.

Reference

4,7-Dimethoxy-5-methyl-coumarin (self-prepared, purity > 99%).

Reagents

Methanol (chromatographic-grade); water (purified water).

Preparation of Samples

10 g of powder sample that had been passed through a 20 mesh sieve was accurately weighed and placed in a 250 ml Soxhlet extractor. 200 ml of chloroform was added to the extractor. After refluxing for 6 h at 95°C, the sample solution was filtered. The chloroform was recovered

from the filtrate until the filtrate was dried out. The residue was dissolved in methanol and topped up to 10 ml in a volumetric flask. After mixing well, it was ready to be used as a test sample.

Preparation of Reference

The reference was made by dissolving 4,7-dimethoxy-5-methyl-coumarin in methanol to obtain the final concentration of 10 μg/ml.

HPLC Conditions

Instrument: LC-10ATVP high-performance liquid chromatography, Shimadzu SPD-10AVP ultraviolet/visible detector
Mobile phase: Methanol:water (45:55)
Flow rate: 0.8 ml/min
Column: Kromasil C_{18} (200mm × 4.6mm, 5μm), Dalian Elite Analytical Instruments Co., Ltd.
Column temperature: Room temperature
Detecting wavelength: 323 nm
Run time: 65 min
Injection volume: 10 μl sample, 10 μl reference

Methods

The sample solution was injected into the liquid chromatograph. The chromatogram was obtained under selected HPLC conditions. The experimental method was validated.

Mobile Phase Selection

A comparison of the experimental results showed that methanol: water (45:55) could successfully separate the peaks and produce the maximum number of peaks from the samples. The duration of the analysis was 65 min. The 1.5 h chromatogram of the samples showed no characteristic peaks after 65 min.

Method Validation

1. Stability: The *Rhizoma imperatae* originating from Inner Mongolia Province was used as test samples to investigate the 24 h stability of the solution. The relative retention times of the common peaks and the relative peak areas that are equal to or more than 10% of the total peak area were analyzed. The RSD was less than 3%, which indicated that the test samples were stable within 24 h.

2. Precision: The *Rhizoma imperatae* originating from Inner Mongolia Province was used as test samples. Five consecutive injections of the samples were conducted to obtain the fingerprints. The relative retention times of the common peaks and the individual peak areas that are equal to or more than 10% of the total peak area were analyzed. The RSD was less than 3%, which indicated that the injections of the test samples were carried out with precision.

3. Reproducibility: The *Rhizoma imperatae* originating from Inner Mongolia was used as test samples ($n = 5$). The samples were prepared according to the preparation and analysis methods for obtaining test sample fingerprints and an HPLC analysis was performed according to the protocol for fingerprint reproducibility. The relative retention times of the common peaks and the individual peak areas that are equal to or more than 10% of the total peak area were analyzed. The results indicated good reproducibility of the results.

Sample Analysis

The HPLC fingerprints of the samples from 10 batches were analyzed under the conditions described above.

Establishment of Fingerprints

The number of peaks, peak area and relative retention time of 10 batches of samples were analyzed and compared to obtain an optimized fingerprint (Fig. 13.1).

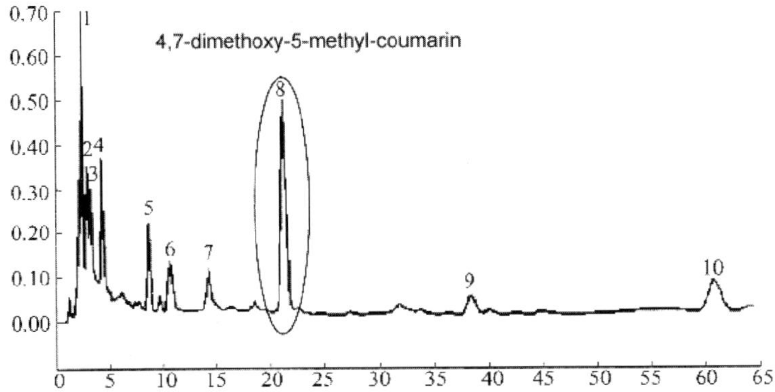

Fig. 13.1. HPLC fingerprint of Baimaogen (*Rhizoma imperatae*).

Results and Discussion

About 20 peaks were separated by HPLC. After comparing all the tested and recorded fingerprints with reference controls, 10 common peaks were selected as the characteristic peaks of the *Rhizoma imperatae* fingerprint.

Common Peak Standardization

Common peaks were assigned according to the results of 10 batches of samples. There were 10 common peaks in the *Rhizoma imperatae* HPLC fingerprints. Compared to the 4,7-dimethoxy-5-methyl-coumarin reference peak (peak 8), the relative retention times of common peaks 1, 2, 3, 4, 5, 9 and 10 were 0.124, 0.149, 0.166, 0.215, 0.417, 1.794 and 2.870, respectively.

Ratio of Common Peak Areas

Aside from the reference peak area, none of the common peak areas exceeded 10% of the total peak area.

2.14

Baishouwu

白首乌 (Bunge Auriculate Root)
Radix Cynanchi auriculati
Radix Cynanchi wilfordii
Radix Cynanchi bungei

Introduction

Bunge Auriculate Root is the dried tuberous root of *Cynanchum auriculatum* Royle ex Wight, *Cynanchum wilfordii* (Maxim.) Hemsl. or *Cynanchum bungei* Decne. (family Asclepiadaceae). *Cynanchum auriculatum* Royle ex Wight mainly originates from Jiangsu Province. *Cynanchum wilfordii* (Maxim.) Hemsl. mainly originates from Jilin and Shandong Provinces. *Cynanchum bungei* Decne. mainly originates from Shandong Province. They are mainly cultivated species.

According to studies on resources, Bunge Auriculate Root (*Cynanchum auriculatum* Royle ex Wight) is planted abundantly in Binhai, Xiangshui and Sheyang of Jiangsu Province. Binhai is famous for its Bunge Auriculate Root. It has a long history of Bunge Auriculate Root cultivation with high yield and high quality. The Bunge Auriculate Root produced there is sold all over the country and is made into refined Heshouwu powder to be sold as a nutritional supplement. (The commercial name of the refined Heshouwu powder is Baishowu Powder.)

Chemical Composition

Bunge Auriculate Root contains various C_{21} steroidal glycosides. Wilfosides C_3N, C_1N, C_1G, K_1N, cynauricuosides A, B, C, and four types of aglucons (gagamine, caudatin, metaplexigenin and kidjoranin) were isolated from *Cynanchum auriculatum* Royle ex Wight. Bunge Auriculate Root also contains phosphatides, which are mainly phosphatidylcholine, phosphatidylethanolamine and phosphatidylinositos.

Experimental Materials

Samples

Bunge Auriculate Root was collected from cultivated species in Wuxun and Zhendong in Binhai of Jiangsu Province in December 2001. There were 15 samples with variant identifications completed on all of them.

Reagents

Acetonitrile (chromatographic-grade, Tedia Company Inc., USA). Double-distilled deionized water (prepared by our lab). Ethanol (analytical-grade methanol was purchased from Shanghai Chemical Reagents Pte. Ltd.; chromatographic-grade methanol was purchased from Tedia Company Inc.). Chloroform (analytical-grade, Shanghai Chemical Reagents Pte. Ltd.). Isopropanol (analytical-grade, Shanghai Chemical Reagents Pte. Ltd.).

Preparation of Samples

5 g of crude powder of the sample was accurately weighed and placed in a round-bottom flask. 100 ml of ethanol was added to the flask. After refluxing three times, the filtrates were combined and further concentrated into a small quantity. It was then extracted twice with 40 ml

of chloroform. The extracts were combined, dehydrated by anhydrous sodium sulfate, and filtered. The filtrates were concentrated to a reasonable volume and evaporated to dryness. The residue was dissolved in isopropanol and topped up to 25 ml in a volumetric flask. After being filtered through a 0.45 μm Millipore filter, it was then ready to be used as a test sample.

HPLC Conditions

Instrument: Agilent 1100 high-performance liquid chromatography and diode array detector
Mobile phase: Water and acetonitrile with gradient elution
Flow rate: 1.0 ml/min
Column: Agilent ZORBAX 80A Extend-C_{18} analytical column (250 mm × 4.6 mm, 5 μm)
Column temperature: 25°C
Detecting wavelength: 210 nm
Reference wavelength: 360 nm
Run time: 80 min
Injection volume: 10 μl

Methods

The sample solution was injected into the liquid chromatograph. The chromatogram was obtained under selected HPLC conditions. The experimental method was validated.

Mobile Phase Selection

A comparison of experimental results showed that gradient elution with water and acetonitrile could successfully separate the peaks and produce the maximum number of peaks from the samples. The duration of the analysis was 80 min. The 2 h chromatogram of the samples showed no characteristic peaks after 80 min.

A is water, B is acetonitrile.

	A (%)	B (%)
0–10 min	75–98	2–25
10–50 min	35–75	25–65
50–70 min	0–35	65–100
70–80 min	0–0	100–100

Method Validation

1. Stability: Test samples from the same batch were stored at room temperature and injected at regular intervals for testing. The degrees of similarity of the injections calculated by the Similarity Degree Calculation Software for Traditional Chinese Medicine Fingerprints were all above 0.94, which indicated that samples were stable within the 36 testing hours.
2. Precision: Five consecutive injections of the samples from the same batch were carried out to determine the precision. The degrees of similarity of the injections calculated by the Similarity Degree Calculation Software for Traditional Chinese Medicine Fingerprints were all above 0.94, which indicated that the injections of the test samples were carried out with precision.
3. Reproducibility: Five samples from the same batch were prepared and tested according to the method described above. The degrees of similarity of the injections calculated by the Similarity Degree Calculation Software for Traditional Chinese Medicine Fingerprints were all above 0.93, indicating good reproducibility of the results.

Sample Analysis

The HPLC fingerprints of the test samples from 10 batches were analyzed under the conditions described above.

Establishment of Fingerprints

The number of peaks, peak area and relative retention time of 10 batches of samples were analyzed and compared to obtain an optimized fingerprint (Figs. 14.1, 14.2).

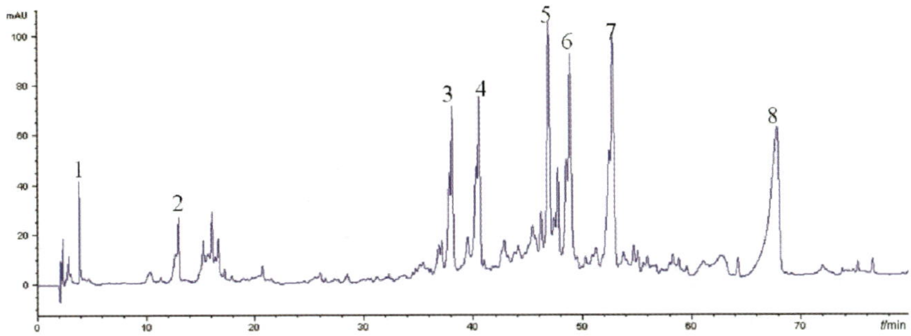

Fig. 14.1. HPLC fingerprint of Baishouwu (Bunge Auriculate Root).

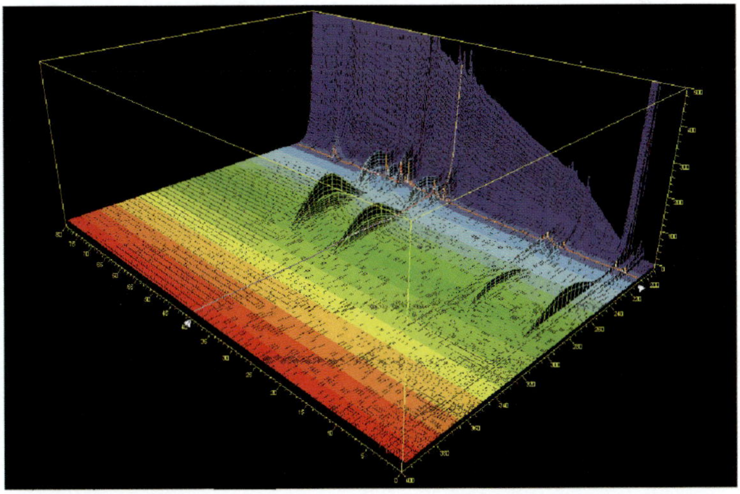

Fig. 14.2. 3-D HPLC fingerprint of Baishouwu (Bunge Auriculate Root).

Results and Discussion

About 40 peaks were separated using HPLC. After comparing all the tested and recorded fingerprints, eight common peaks were selected as the characteristic peaks of the Bunge Auriculate Root fingerprint. The retention time (min) and peak area values of the eight common peaks are listed below.

Peak number	1	2	3	4	5	6	7	8
Retention time	3.804	12.906	38.023	40.464	46.840	48.826	52.730	67.774
Peak area	195.8	502.3	1656.7	2047.7	2134.5	2784.9	3069.4	3734.9

Common Peak Standardization

Common peaks were assigned according to the results of 10 batches of samples. There were eight common peaks in the HPLC fingerprints of Bunge Auriculate Root. The common peak with a retention time of about 38 min was subsequently used as the reference peak. Compared to the reference peak (peak 3), the relative retention times of the other seven common peaks were 0.100, 0.339, 1.064, 1.232, 1.284, 1.387 and 1.782.

Ratio of Common Peak Area

The peak area of common peak 8 was more than 10% of the total peak area. The relative peak area ratio of peak 8 was specified as 1.5781–2.9307.

2.15

Dihuang

地黄

Radix rehmanniae

Introduction

Radix rehmanniae is the fresh or dried tuberous root of *Rehmannia glutinosa* Libosch. (family Scrophulariaceae). It mainly originates from Henan Province. It is usually called Huai Dihuang (怀地黄) and most of the plants are cultivated species.

According to studies and investigations on resources, Huai Dihuang is planted abundantly in Wuzhi, Wenxian, Mengxian, Bo'ai, Jiaozuo, Xiuwu, and Qinyang (previously Huai Qingfu) of Henan Province. Huai Dihuang planted in these areas has the highest quality and abundance. Not only is it sold in China, it is also exported to other countries in large quantities.

Chemical Composition

Radix rehmanniae contains iridoids and their glycosides, which mainly are rehmanniosides A, B, C, D, melittoside, aucubin, catalpol, rehmannin and campesterol.

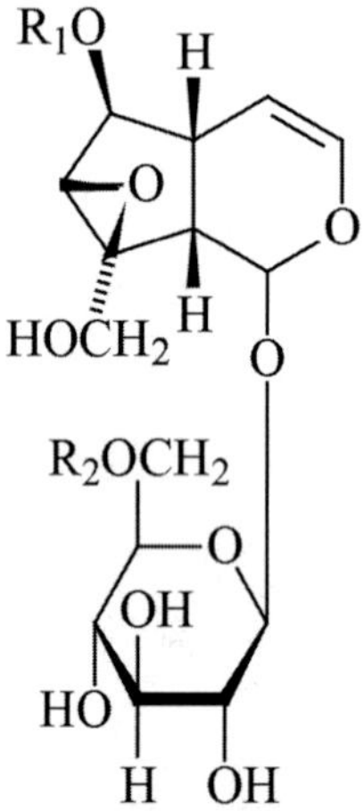

Rehmannioside A $R_1=\alpha$**-D-gal*,** $R_2=H$
Rehmannioside B $R_1=H$**,** $R_2=\alpha$**-D-gal**

***gal is galactose, *sic passim*.**

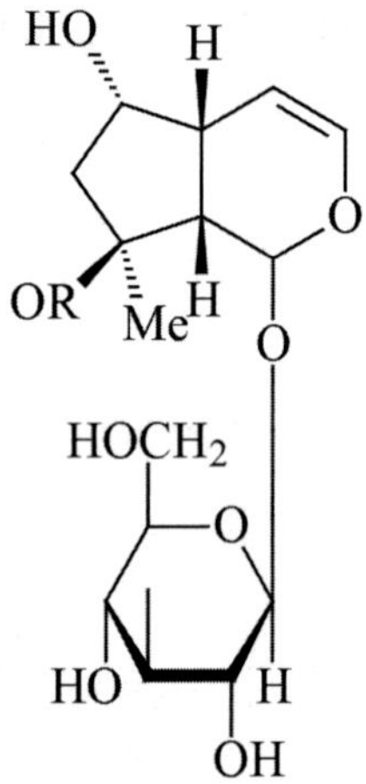

Rehmannioside C R=α**-D-gal**

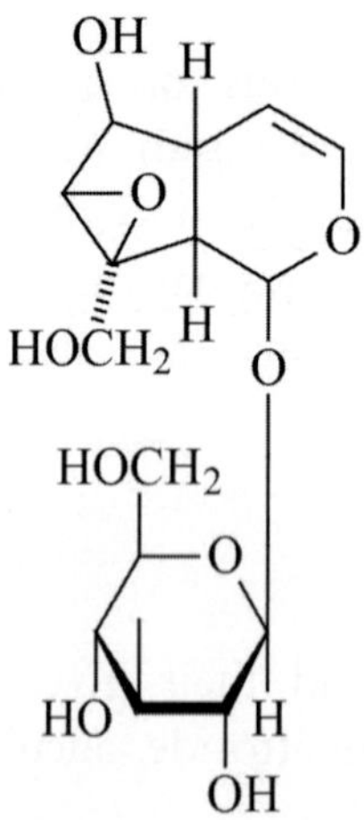

Rehmannioside D R=D-glu$^{1\text{-}2}$**O-**β**-D-glu**
Melittoside R=O-β**-D-glu**

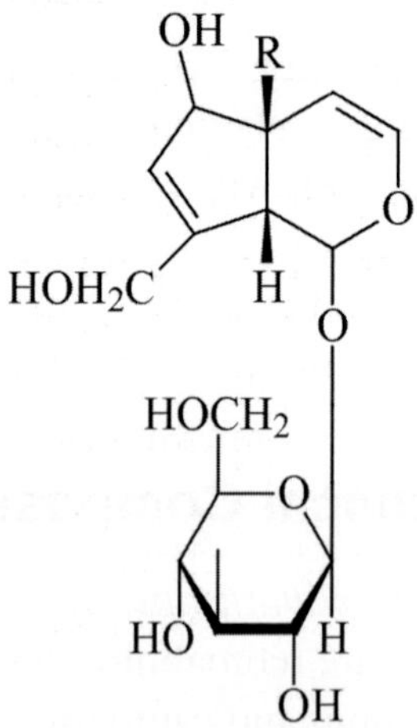

Catalpol

Experimental Materials

Samples

Radix rehmanniae were collected from Wuzhi, Jiaozuo and Qinyang of Henan Province in October 2002. There were 14 samples with variant identifications completed on all of them.

Reagents

Methanol (analytical-grade methanol was purchased from Shanghai Chemical Reagents Pte. Ltd.; chromatographic-grade methanol was purchased from Tedia Company Inc.). *n*-Butanol (analytical-grade, Shanghai Chemical Reagents Pte. Ltd.). Double-distilled deionized water (prepared by our lab). Phosphoric acid (analytical-grade).

Preparation of Samples

10 g of crude powder of the sample was accurately weighed and placed in a round-bottom flask. It was dissolved 1:8 in water. After refluxing for 2 h, the extract was concentrated to 50 ml and placed in a 125 ml separating funnel. 50 ml of *n*-butanol was added to the funnel. After being extracted with shaking, it was set aside for the layers to separate. The layer of *n*-butanol was collected, placed in an evaporating dish and dried over a 100°C water bath. The residue was dissolved in a proper amount of methanol and filtered through a 0.45 μm Millipore filter. After being topped up to 10 ml, it was then ready to be used as a test sample.

HPLC Conditions

Instrument: Agilent 1100 high-performance liquid chromatography and diode array detector
Mobile phase: Methanol and 0.01% phosphoric acid with two-step gradient elution

Flow rate: 1.0 ml/min
Column: HYPERSIL C_{18} analytical column (250 mm × 4.6 mm, 10 μm)
Column temperature: 30°C
Detecting wavelength: 236 nm
Reference wavelength: 360 nm
Run time: 40 min
Injection volume: 10 μl

Methods

The sample solution was injected into the liquid chromatograph. The chromatogram was obtained under selected HPLC conditions. The experimental method was validated.

Mobile Phase Selection

A comparison of the experimental results showed that two-step gradient elution with the methanol and 0.01% phosphoric acid solution could successfully separate the chromatographic peaks and produce the maximum number of peaks from the samples. The duration of the analysis was 40 min. The 1.5 h chromatogram of the samples showed no characteristic peaks after 40 min.

A is methanol, B is 0.01% phosphoric acid.

	A (%)	B (%)
0–15 min	5–30	70–95
15–45 min	30–75	25–70
45–55 min	75–75	25–25

Method Validation

1. Stability: Test samples were stored at room temperature and analyzed at 0 h, 1 h, 2 h, 4 h, 6 h, 24 h and 48 h. The degrees of similarity were analyzed by the Similarity Degree Calculation

Software for Traditional Chinese Medicine Fingerprints. The results showed that the degrees of similarity were all above 0.9, which indicated that samples were stable within the 48 testing hours.

2. Precision: Five consecutive injections of the samples were carried out to determine the precision. The degrees of similarity were analyzed the Similarity Degree Calculation Software for Traditional Chinese Medicine Fingerprints. The results showed that the degrees of similarity were all above 0.9, which indicated that the injections of the test samples were carried out with precision.
3. Reproducibility: Five samples from the same batch were prepared and tested simultaneously according to the method described above. The degrees of similarity were analyzed by the Similarity Degree Calculation Software for Traditional Chinese Medicine Fingerprints. The results showed that the degrees of similarity were all above 0.9, indicating good reproducibility of the results.

Establishment of Fingerprints

The number of peaks, peak area and relative retention time of 10 batches of samples were analyzed and compared to obtain an optimized fingerprint (Fig. 15.1).

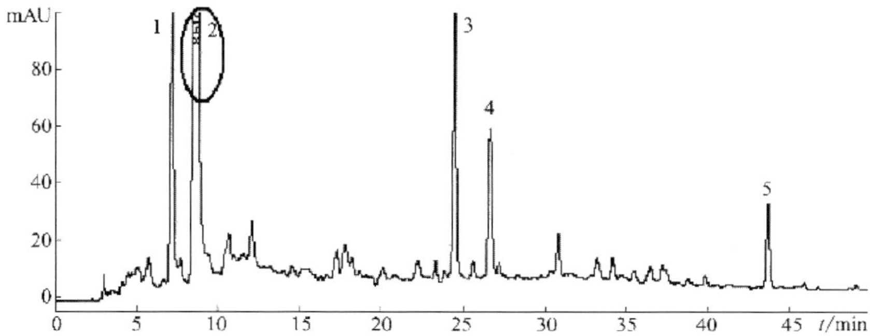

Fig. 15.1. HPLC fingerprint of Dihuang (*Radix rehmanniae*).

Results and Discussion

More than 30 peaks were separated using HPLC. After comparing all the tested and recorded fingerprints, five common peaks were selected to be the characteristic peaks of the *Radix rehmanniae* fingerprint. The retention time (min) and peak area values of the five common peaks are listed below.

Peak number	1	2	3	4	5
Retention time	7.256	8.896	24.672	26.839	43.910
Peak area	1758.24	15939.30	1839.19	1083.21	548.96

Common Peak Standardization

Common peaks were assigned according to the results of 10 batches of samples. There were five common peaks in the HPLC fingerprints of *Radix rehmanniae*. The common peak with a retention time of about 8.896 min was subsequently used as the reference peak. Compared to the reference peak (peak 2), the relative retention times of the other four common peaks were 0.816, 2.773, 3.017 and 4.936.

Characteristics of Common Peaks

The typical characteristic of the common peaks was that peak 2 was the strongest peak, while peaks 1 and 3 were moderate peaks. The peak areas decreased progressively from peak 3 to peak 4, and then to peak 5.

Ratio of Common Peak Area

Aside from the reference peak area, none of the common peak areas exceeded 10% of the total peak area.

2.16

Xiqingguo
西青果
Fructus chebulae immaturus

Introduction

Fructus chebulae immaturus is the dried immature fruit of *Terminalia chebula* Retz. or *Terminalia chebula* Retz. var. *tomentella* Kurt. (family Combretaceae).

The species of *Fructus chebulae immaturus* documented in Quality Standards for Imported Medicines (Ministry of Health, China) is the immature fruit of *Terminalia chebula* Retz. The literature shows that the HPLC fingerprints of 11 imported *Fructus chebulae immaturus* samples, 10 commercial samples, and samples from different origins have been established. Except for two samples, all other *Fructus chebulae immaturus* samples have similar HPLC fingerprints. After comparing it with the HPLC fingerprints of different types of *Terminalia chebulae*, it was found that the fingerprint of *Fructus chebulae immaturus* was very similar to that of *Terminalia chebula* Retz. f. *macrocarpa*. This indicated that the origin of part of the collected samples was very likely to be the immature fruit of *Terminalia chebula* Retz. f. *macrocarpa*. Most of the samples were *Terminalia chebula* Retz., which was the same as the one documented in Import Medicinal Material Rules.

Traditionally, *Fructus chebulae immaturus* is imported from India and Sri Lanka. Recently, there have been some supplied from Yunnan Province, etc, and their species have not been identified.

Chemical Composition

Fructus chebulae immaturus contains tannins, triterpenoids, phenolic acids and aliphatic compounds. Some tannins, such as gallic acid, chebulagic acid and chebulinic acid, are the main chemical components of *Fructus chebulae*, which contribute 23–37% of the chemical composition.

Experimental Materials

Samples

Forty-two *Fructus chebulae immaturus* samples were collected from different markets (including imported samples). Most of the samples were *Terminalia chebula* Retz., which was the same as the one documented in Quality Standards for Imported Medicines.

References

Gallic acid, chebulagic acid and chebulinic acid for quantitative analysis were provided by the National Institute for the Control of Pharmaceutical and Biological Products (China).

Reagents

Ethyl acetate and methanol (analytical-grade), acetone (analytical-grade), methanol (analytical-grade), phosphoric acid (analytical-grade), potassium dihydrogen phosphate (analytical-grade), double-distilled deionized water (prepared by our lab).

Preparation of Samples

0.05 g of powder sample (passed through a 3 mesh sieve) was refluxed twice with 50 ml of acetone for 30 min each time. It was then filtered and the filtrate was evaporated to dryness. The residue was topped up to 10 ml with water and placed in a fridge (5–10°C) for about 1 h. It was then filtered with a 0.5 μm Millipore filter and the filtrate was ready to be used as a test sample.

Preparation of References

Gallic acid, chebulagic acid, and chebulinic acid controls were prepared as 50 mg/ml aqueous solutions. After being filtered through a 0.45 μm Millipore filter, they were then ready to be used as references.

HPLC Conditions

Instrument: Agilent 1100 high-performance liquid chromatography and diode array detector
Mobile phase: Three-step gradient elution with 0.05 mol/L phosphoric acid and 0.05 mol/L potassium dihydrogen phosphate (1:1), methanol and ethyl acetate
Flow rate: 1.0 ml/min
Column: LiChrospher analytical column RP-18 (4 mm × 125 mm, 5 mm); precolumn is Phenomenex C_{18}
Column temperature: 20°C
Detecting wavelength: 280 nm
Run time: 40 min
Injection volume: 20 μl sample, 20 μl reference

Methods

The sample solution was injected into the liquid chromatograph. The chromatogram was obtained under selected HPLC conditions. The experimental method was validated.

Mobile Phase Selection

A comparison of the experimental results showed that three-step gradient elution with 0.05 mol/L phosphoric acid and 0.05 mol/L potassium dihydrogen phosphate (1:1), methanol and ethyl acetate could successfully separate the chromatographic peaks and produce the maximum number of peaks from the samples. The duration of the analysis was 40 min. The 1.5 h chromatogram of the samples showed no characteristic peaks after 40 min.

A is 0.05 mol/L phosphoric acid and 0.05 mol/L potassium dihydrogen phosphate (1:1); B is methanol; C is ethyl acetate.

	A (%)	B (%)	C (%)
0–5 min	94–96	3–6	0–1
5–15 min	93–96	2–3	1–5
15–20 min	89–93	2–6	5–5
20–35 min	55–89	6–40	5–5

Method Validation

1. Stability: Test samples were stored at room temperature and tested at 0 h, 1 h, 2 h, 3 h, 4 h, 5 h, 24 h and 48 h. The degrees of similarity were analyzed by the Similarity Degree Calculation Software for Traditional Chinese Medicine Fingerprints. The results showed that the degrees of similarity were 0.9823–0.9993, which indicated that samples were stable within the 48 testing hours.

2. Precision: Five consecutive injections of the samples were carried out to determine the precision. The degrees of similarity were analyzed by the Similarity Degree Calculation Software for Traditional Chinese Medicine Fingerprints. The results showed that the degrees of similarity were 0.9806–0.9992, which indicated that the injections of the test samples were carried out with precision.

3. Reproducibility: Five samples from the same batch were prepared and tested according to the method described above. The degrees of similarity were analyzed by the Similarity Degree Calculation Software for Traditional Chinese Medicine Fingerprints. The results showed that the degrees of similarity were all above 0.9929, indicating good reproducibility of the results.

Sample Analysis

The HPLC fingerprints of the test samples from 10 batches were analyzed under the conditions described above.

Establishment of Fingerprints

The number of peaks, peak area and relative retention time of 10 batches of samples were analyzed and compared to obtain an optimized fingerprint (Figs. 16.1, 16.2).

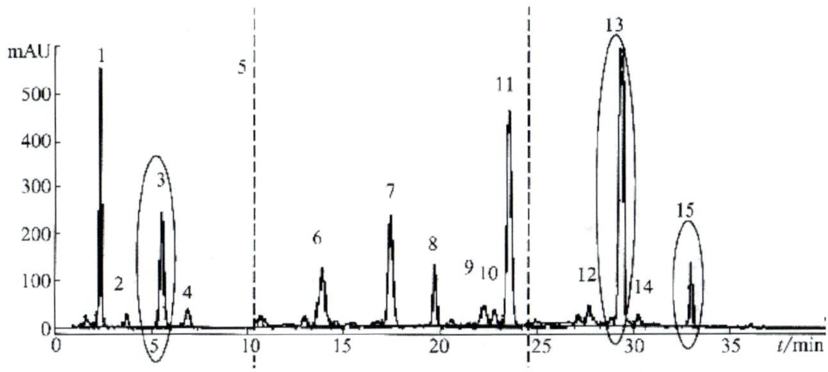

Note: Gallic acid t_R = 5.558 min, chebulagic acid t_R = 29.342 min, and chebulinic acid t_R = 32.994 min.

Fig. 16.1. HPLC fingerprint of Xiqingguo (*Fructus chebulae immaturus*).

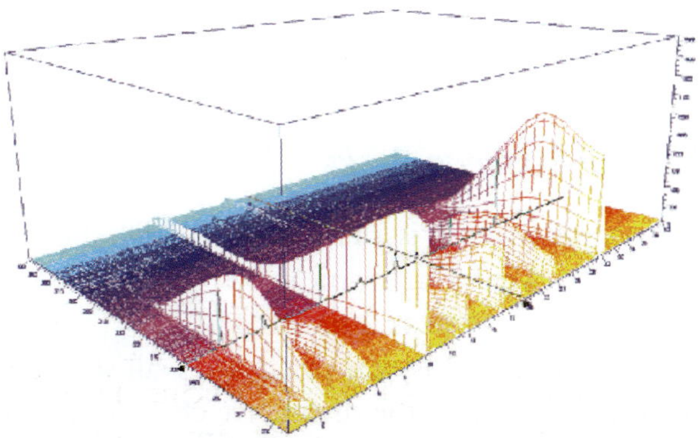

Fig. 16.2. 3-D HPLC fingerprint of Xiqingguo (*Fructus chebulae immaturus*).

Results and Discussion

More than 20 peaks were separated using HPLC. After comparing all the tested and documented fingerprints, 15 common peaks were selected as the characteristic peaks of the *Fructus chebulae immaturus* fingerprint. The retention time (min) and peak area values of the 15 common peaks are listed below.

Peak number	1	2	3	4	5	6	7	8
Retention time	2.356	3.857	5.561	6.975	10.137	13.974	17.521	19.836
Peak area	2575.29	294.28	6493.52	419.80	1667.14	2823.44	6752.56	2188.91

Peak number	9	10	11	12	13	14	15
Retention time	22.375	22.938	23.789	27.862	29.427	30.241	33.013
Peak area	1352.94	1152.27	6864.80	2181.89	19985.3	892.60	2163.06

Common Peak Standardization

Common peaks were assigned according to the results of 10 batches of samples. There were 15 common peaks in the HPLC fingerprints of *Fructus chebulae immaturus*. Compared to the chebulagic acid reference peak (peak 13), the relative retention times of the other 14 common peaks were 0.080, 0.131, 0.189, 0.237, 0.344, 0.475, 0.595, 0.674, 0.760, 0.779, 0.808, 0.947, 1.028 and 1.122.

Characteristics of Common Peaks

The characteristic of zone 1 is that peaks 1–3 all showed up. The strengths of the peaks were roughly measured through ordinate marks. It was found that peak 1 >>peak 2 (gallic acid) <<<peak 3. The common characteristic of zone 2 is that peaks 4–11 all showed up. Peaks 8–10 were weaker, which was the distinctive characteristic of *Fructus*

Chebulae Immaturus. There were only four characteristic peaks in zone 3. Peak 13 <<<peak 15 >>>peak 16 <<peak 19.

Ratio of Common Peak Area

Aside from the chebulagic acid reference peak area, the peak areas of common peaks 7 and 11 exceeded 10% of the total peak area. The relative peak area ratio of peak 7 was specified as 0.237–0.439. The relative peak area ratio of peak 11 was specified as 0.240–0.447.

2.17

Honghua
红花
Flos carthami

Introduction

Flos carthami is the dried flower of *Carthamus tinctorius* L. (family Compositae). It mainly originates from Henan, Anhui, Sichuan, Jiangsu and Zhejiang Provinces.

According to studies on resources, *Flos carthami* is mainly produced in Xinjiang Province, which contributes 80% of the country's total production. It is planted abundantly in Jimusaer, Qitai, Yanqi, Yingjisha, Muleihasake, Hutubi, Hejing, Changji, Yutian, Shache, Huocheng, Shule, Chabuchaer, Yumin, Tacheng, Kuche and Shaya of Xinjiang Province. Jimusaer Red Flag Farm is the largest *Flos carthami* planting base in China and is renowned for its excellent and stable quality of *Flos carthami*.

Chemical Composition

Flos carthami contains carthamin, carthamone and neocarthamin. The chemical composition of *Flos carthami* in different mature stages is different. The canary flower mainly contains neocarthamin and small amounts of carthamin; the yellow flower mainly contains carthamin; the orange flower mainly contains carthamin or carthamone. *Flos carthami* also contains carthamidin, safflower yellow, nonacosane, β-sitosterol, palmitic acid, myristic acid and lauric acid.

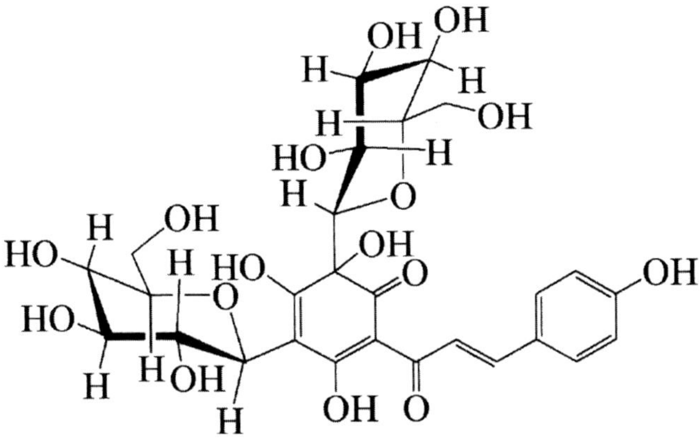

Hydroxyl safflower yellow

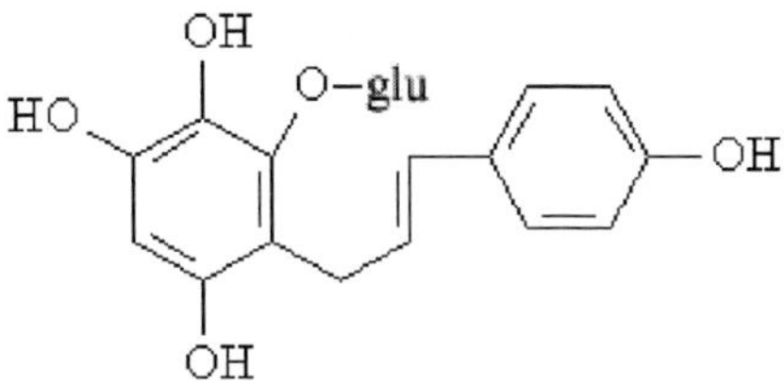

Carthamin

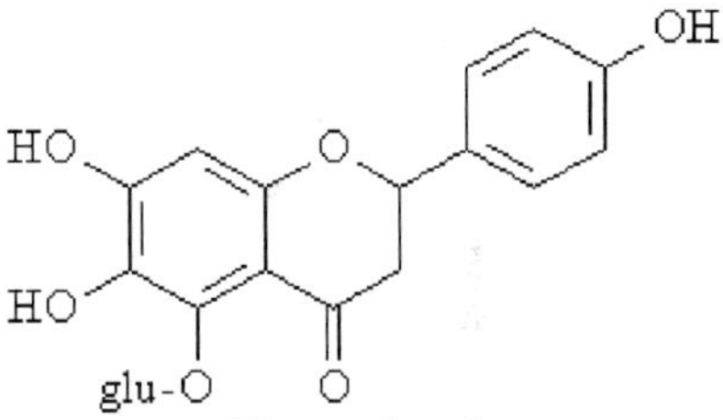

Neocarthamin

Experimental Materials

Samples

Flos carthami samples were collected from Yanqi, Qitai, Yingjisha, Mulei, Yutian, Jimusaer and Hejing of Xinjiang Province, Baotou of Inner Mongolia Province, Xi'an of Shaanxi Province, Xinxiang of Henan Province, and Chuzhou of Anhui Province. There were 31 samples with variant identifications completed on all of them.

Reference

Hydroxyl safflower yellow A was prepared by our lab.

Reagents

Acetonitrile (chromatographic-grade, Tedia Company Inc., USA). Double-distilled deionized water (prepared by our lab). Phosphoric acid (analytical-grade, Shanghai Chemical Reagents Pte. Ltd.).

Preparation of Samples

2 g of powder sample (passed through a 4 mesh sieve) was accurately weighed and placed in a round-bottom flask. After adding 50 ml of water, it was weighed again. It was then extracted in a 90°C water bath for 1 h and weighed. The lost weight was topped up and the solution was filtered. 10 ml of filtrate was accurately measured. After being further filtered with a Millipore filter, it was then ready to be used as a test sample.

Preparation of Reference

An appropriate amount of vacuum-dried hydroxyl safflower yellow A was accurately weighed and dissolved in water to obtain a concentration of 0.2 g/L. It was then ready to be used as a reference.

HPLC Conditions

Instrument: Waters 600 high-performance liquid chromatography, diode array detector 2996, and Waters Millennium workstation
Mobile phase: Gradient elution with acetonitrile and 0.5% phosphoric acid
Flow rate: 1.0 ml/min
Column: Phenomenex Luna (250 mm × 4.6 mm, 5 μm)
Column temperature: 25°C
Detecting wavelength: 275 nm
Run time: 65 min
Injection volume: 10 μl sample, 10 μl reference

Methods

The sample solution was injected into the liquid chromatograph. The chromatogram was obtained under selected HPLC conditions. The experimental method was validated.

Mobile Phase Selection

A comparison of the experimental results showed that gradient elution with acetonitrile and 0.5% phosphoric acid could successfully separate the chromatographic peaks and produce the maximum number of peaks from the samples. The duration of the analysis was 65 min. The 1.5 h chromatogram of the samples showed no characteristic peaks after 65 min.

A is acetonitrile, B is 0.5% phosphoric acid.

	A (%)	B (%)
0–55 min	0–25	75–100
55–65 min	0–25	75–100

Method Validation

1. Stability: Test samples from the same batch were stored at room temperature and injected at regular intervals for testing. The degrees of similarity of the injections calculated by the Similarity Degree Calculation Software for Traditional Chinese Medicine Fingerprints were all above 0.95, which indicated that samples were stable within the 24 testing hours.
2. Precision: Five consecutive injections of the samples from the same batch were carried out to determine the precision. The degrees of similarity of the injections calculated by the Similarity Degree Calculation Software for Traditional Chinese Medicine Fingerprints were all above 0.95, which indicated that the injections of the test samples were carried out with precision.
3. Reproducibility: Five samples from the same batch were prepared and tested according to the method described above. The degrees

of the similarity of the injections calculated by the Similarity Degree Calculation Software for Traditional Chinese Medicine Fingerprints were all above 0.95, indicating good reproducibility of the results.

Sample Analysis

The HPLC fingerprints of the test samples from 10 batches were analyzed under the conditions described above.

Establishment of Fingerprints

The number of peaks, peak area and relative retention time of 31 batches of samples were analyzed and compared to obtain an optimized fingerprint (Fig. 17.1).

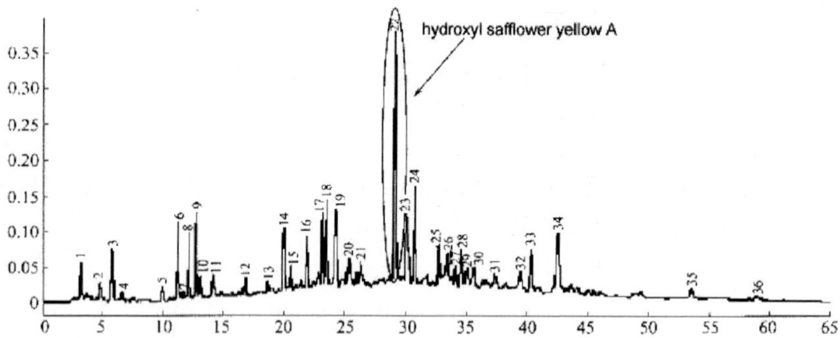

Fig. 17.1. HPLC fingerprint of Honghua (*Flos carthami*).

Results and Discussion

About 50 peaks were separated using HPLC. After comparing all the tested and documented fingerprints, 36 common peaks were selected as the characteristic peaks of *Flos carthami*. The retention time (min) and peak area values of the 36 common peaks are listed below.

Peak number	1	2	3	4	5	6	7	8	9
Retention time	3.27	4.90	5.87	6.68	10.01	11.32	11.80	12.23	12.83

Peak area	425108	325416	666665	162136	222568	730921	169233	769156	922034
Peak number	10	11	12	13	14	15	16	17	18
Retention time	13.18	14.22	16.81	18.66	20.02	20.55	21.91	23.20	23.50
Peak area	425586	488722	330665	329674	1447342	791308	1091393	1128666	1488162
Peak number	19	20	21	22	23	24	25	26	27
Retention time	24.30	25.42	26.32	29.12	30.11	30.76	32.77	33.48	34.16
Peak area	2083063	1016956	1284903	4944965	2219283	1711490	807365	1208226	758760
Peak number	28	29	30	31	32	33	34	35	36
Retention time	34.62	35.11	35.65	37.45	39.54	40.39	42.58	53.52	59.01
Peak area	991910	871849	901770	979308	1069211	982786	1860576	243628	139046

Common Peak Standardization

Common peaks were assigned according to the results of 10 batches of samples. There were 36 common peaks in the HPLC fingerprints of *Flos carthami*. Compared to the hydroxyl safflower yellow A reference peak (peak 22), the relative retention times of the other 35 common peaks were 0.11, 0.17, 0.20, 0.23, 0.34, 0.39, 0.41, 0.42, 0.44, 0.45, 0.49, 0.58, 0.64, 0.69, 0.71, 0.75, 0.80, 0.81, 0.83, 0.87, 0.90, 1.03, 1.06, 1.13, 1.15, 1.17, 1.19, 1.21, 1.22, 1.29, 1.36, 1.39, 1.46, 1.84 and 2.03.

Ratio of Common Peak Area

Aside from the reference peak, none of the common peak areas exceeded 10% of the total peak area.

Similarity Degree Evaluation

The HPLC fingerprints of *Flos carthami* bought from different places were documented and analyzed by the Computer-assisted Similarity Degree Evaluation System. The degrees of similarity samples a–n were 0.96, 0.78, 0.88, 0.96, 0.91, 0.95, 0.87, 0.93, 0.96, 0.96, 0.96, 0.91,

0.94 and 0.93, respectively.

After scrutinizing the fingerprints of 14 batches of commercial *Flos carthami*, it can be seen that the differences lie mainly in the peak shapes and relative heights of peaks 4–6.

In the fingerprints of *Flos carthami* from Xi'an of Shaanxi Province and Baotou of Inner Mongolia Province, peak 5 was relatively higher. Usually, peak 22 (hydroxyl safflower yellow A) was the highest peak. But peak 5 of *Flos carthami* from these two places became the highest peak, which lowered the degrees of similarity. The degrees of similarity were 0.78 and 0.87, respectively.

For the fingerprints of *Flos carthami* from Huhehaote of Inner Mongolia Province, peaks 4 and 5 were higher. The degree of similarity of 0.88 was relatively low. For the fingerprints of *Flos carthami* from Xinxiang Wucunzhen and Shaoercun of Henan Province, peak 6 was relatively low. For the fingerprints of *Flos carthami* from Chuzhou of Aihui Province, peak 4 was relatively high.

Xinjiang Province is the main source of *Flos carthami*. To assist in the selection of high-quality medicinal materials for the Chinese medicine industry and for clinical purposes, the fingerprints of *Flos carthami* from different places in Xinjiang Province were systematically investigated. Their degrees of similarity were evaluated to be 0.97, 0.97, 0.98, 0.99, 0.96, 0.96 and 0.95.

A comparison of the fingerprints of *Flos carthami* from different places in Xinjiang Province showed that the differences were mainly in the peak shapes and relative heights of peaks 2, 3, 4, 5, 6, 8 and 9. Peaks 3, 6, 8 and 9 of samples from Yanqi, Qitai and Yingjisha were all very weak and peak 4 of these samples was stronger. Peaks 3 and 6 of samples from Mulei were very weak. Peak 4 of samples from Yutian and Jimusaer was also very weak. Peak 6 of samples from Hejing was very weak.

The fingerprint results of 31 batches of test samples indicated that the shapes and relative proportions of peaks 12–36, which have the relative retention time of 15–65 min, were relatively stable. This is a common characteristic of *Flos carthami*. Peaks 1–11, which have the relative retention time of 0–15 min, demonstrated greater deviations among different samples.

Jimusaer Red Flag Farm (Xinjiang Province) is the largest *Flos Carthami* planting base in China and is renowned for its excellent and stable quality of *Flos carthami*. Currently, most of the raw materials for producing Honghua (*Flos carthami*) injection come from this base. To further investigate the quality of *Flos carthami* produced in this base and to assist the selection of *Flos carthami* for producing injections, 10 batches of representative samples were collected from this base and their HPLC fingerprints were investigated. The degrees of similarity of these 10 batches of *Flos carthami* samples were evaluated as 0.97, 0.95, 0.98, 0.98, 0.98, 0.98, 0.99, 0.98, 0.92 and 0.94.

2.18

Maidong
麦冬
Radix ophiopogonis

Introduction

Radix ophiopogonis is the dried root tuber of *Ophiopogon japonicus* (Thunb.) Ker-Gawl. (family Liliaceae). Those originating from Cixi, Yuyao and Hangzhou of Zhejiang Province are called Hang Maidong (杭麦冬). Those originating from Mianyang of Sichuan Province are called Chuan Maidong (川麦冬).

According to studies on resources, *Radix ophiopogonis* is planted abundantly in Mianyang, Santai, Jiangyou, Nanbu, Shehong and Suining of Sichuan Province. Mianyang and Santai are famous for their Chuan Maidong.

Chemical Composition

The root tuber contains various saponins, such as ophiopognins A, B, B′, C, C′, D, D′, of which the content of ophiopognin A is the highest at 0.05%. Ophiopognin B takes the second place at 0.01%. The percentages of ophiopognin C and D are very low. Aglucons of ophiopognins A, B, C and D are all ruscogenin. Aglucons of ophiopognins B′, C′ and D′ are all diosgenins. *Radix ophiopogonis* also contains flavonoids, such as ophiopogonones A and B, methylophiopogonanone, 6-aldehydo-7-*O*-methyl-isoophiopogonanone, and 6-aldehydoisoophiopogonone A, B. In addition, it contains volatile oil, β-sitosterol, stigmasterol, β-sitosterol-β-L-glucopyranoside,

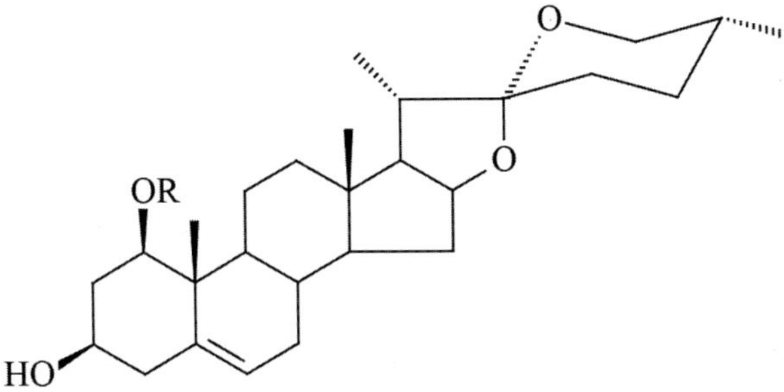

Ruscogenin	**R=H**
Ophiopognin B	**R=L-rha(1–2)-D-fcu*-**
Ophiopognin D	**R=L-rha(1–2)** ⟩ D-fcu- **D-xyl(1–3)**

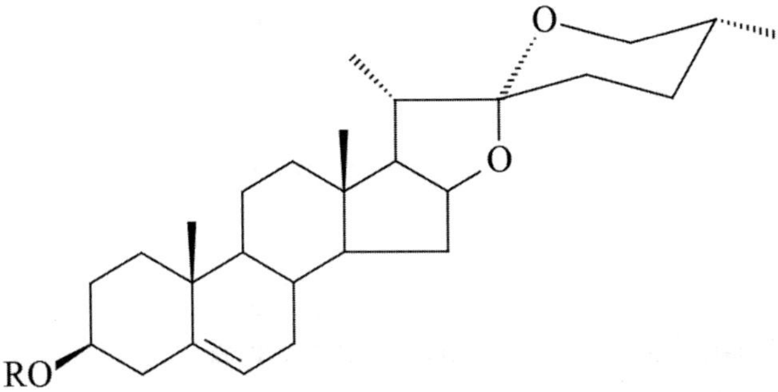

Diosgenin	**R=H**
Ophiopognin B′	R=L-ac-rha(1–2) ⟩ D–glu– D-xyl(1–3)
Ophiopognin C′	**R=L-ac-rha(1–2)-D-glu-**
Ophiopognin D′	R=L-rha(1–2) ⟩ D–glu– D-xyl(1–3)

* fcu is fucose, *sic passim*.

β-sitosterol-3-O-β-D-glucopyranoside and borneol-glucopyranoside. Moreover, it also contains 71% monosaccharides and oligosaccharides. There were over 30 peaks of volatile oil in the gas chromatogram of *Radix ophiopogonis* and 10 of them, such as camphor, linalool, terpineol, α-pinene and β-pinene, were identified. Most of them are monoterpenoids and sesquiterpenoids.

Experimental Materials

Samples

Radix ophiopogonis samples were collected from two-year-old cultivated species in Mianyang of Sichuan Province in April 2002. There were 13 samples with variant identifications completed on all of them.

Reagents

Acetonitrile (chromatographic-grade, Tedia Company Inc., USA). Double-distilled deionized water (prepared by our lab). Methanol (analytical-grade methanol was purchased from Shanghai Chemical Reagents Pte. Ltd.; chromatographic-grade methanol was purchased from Tedia Company Inc.).

Preparation of Samples

2 g of crude powder of the sample was accurately weighed and placed in a round-bottom flask. 100 ml of methanol was added to the flask. After refluxing for 2 h, the sample mixture was filtered. 100 ml of methanol was added to the residue and the sample was extracted for 1 h. The sample mixture was filtered and the filtrates were combined. The filtrate was evaporated to dryness. The residue was mixed with methanol and topped up to 10 ml in a volumetric flask. The mixture was filtered through a 0.45 μm Millipore filter. The filtrate was then ready to be used as a test sample.

HPLC Conditions

Instrument: Agilent 1100 high-performance liquid chromatography and diode array detector
Mobile phase: Water and acetonitrile with gradient elution
Flow rate: 1.0 ml/min
Column: Agilent ZORBAX 80A Extend-C_{18} analytical column (250 mm × 4.6 mm, 5 μm)
Column temperature: 25°C
Detecting wavelength: 230 nm
Reference wavelength: 360 nm
Run time: 40 min
Injection volume: 10 μl

Methods

The sample solution was injected into the liquid chromatograph. The chromatogram was obtained under selected HPLC conditions. The experimental method was validated.

Mobile Phase Selection

A comparison of the experimental results showed that gradient elution with water and acetonitrile could successfully separate the chromatographic peaks and produce the maximum number of peaks from the samples. The duration of the analysis was 40 min. The 1.5 h chromatogram of the samples showed no characteristic peaks after 40 min.

A is water, B is acetonitrile.

	A (%)	B (%)
0–40 min	46–98	2–54

Method Validation

1. Stability: Test samples from the same batch were stored at room temperature and injected at regular intervals for testing. The degrees of similarity of the injections calculated by the Similarity

Degree Calculation Software for Traditional Chinese Medicine Fingerprints were all above 0.92, which indicated that samples were stable within the 36 testing hours.

2. Precision: Five consecutive injections of the samples from the same batch were carried out to determine the precision. The degrees of similarity of the injections calculated by the Similarity Degree Calculation Software for Traditional Chinese Medicine Fingerprints were all above 0.98, which indicated that the injections of the test samples were carried out with precision.
3. Reproducibility: Five samples from the same batch were prepared and tested according to the method described above. The degrees of similarity of the injections calculated by the Similarity Degree Calculation Software for Traditional Chinese Medicine Fingerprints were all above 0.95, indicating good reproducibility of the results.

Sample Analysis

The HPLC fingerprints of the test samples from 10 batches were analyzed under the conditions described above.

Establishment of Fingerprints

The number of peaks, peak area and relative retention time of 10 batches of samples were analyzed and compared to obtain an optimized fingerprint (Figs. 18.1, 18.2).

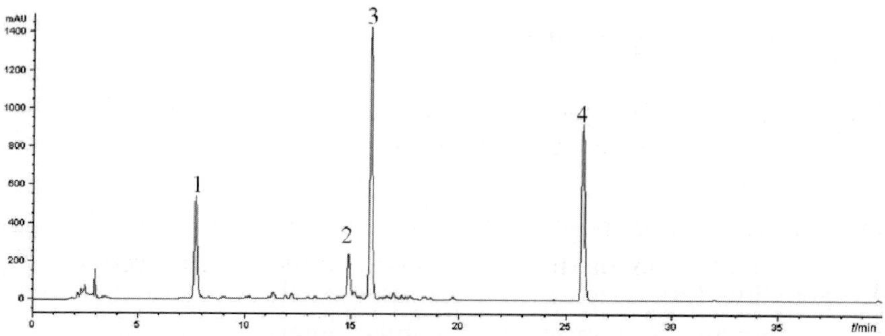

Fig. 18.1. HPLC fingerprint of Maidong (*Radix Ophiopogonis*).

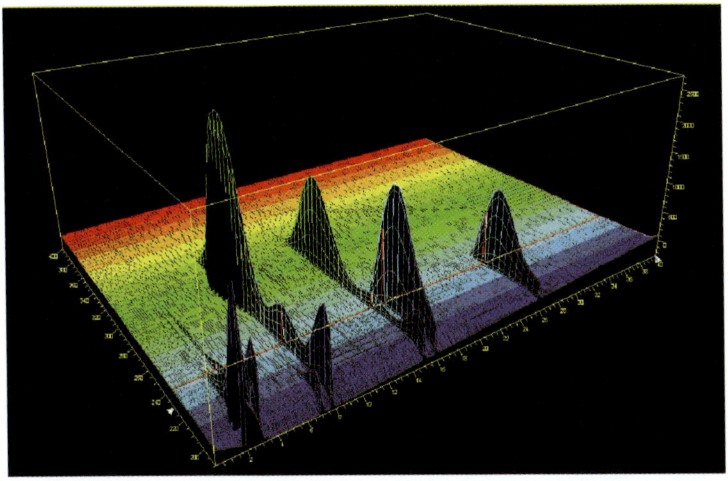

Fig. 18.2. 3-D HPLC fingerprint of Maidong (*Radix ophiopogonis*).

Results and Discussion

More than 10 peaks were separated using HPLC. After comparing all the tested and documented fingerprints, four common peaks were selected as the characteristic peaks of *Radix ophiopogonis*. The retention time (min) and peak area values of the four common peaks are listed below.

Peak number	1	2	3	4
Retention time	7.698	14.861	15.869	25.772
Peak area	5405.6	3211	11754.3	9479.6

Common Peak Standardization

Common peaks were assigned according to the results of 10 batches of samples. There were four common peaks in the HPLC fingerprints of *Radix ophiopogonis*. Common peak 3 from the test samples was used as the reference peak. Compared to the reference peak, the relative retention times of the other three common peaks were 0.485, 0.9036 and 1.624.

Ratio of Common Peak Area

Aside from the reference peak, the peak areas of common peaks 1 and 4 exceeded 10% of the total peak area. The relative peak area ratio of peak 1 was specified as 0.322–0.598. The relative peak area ratio of peak 4 was specified as 0.5642–1.0478.

2.19

Lugen
芦根
Rhizoma phragmitis

Introduction

Rhizoma phragmitis is the fresh or dried rhizome of *Phragmites communis* Trin. (family Gramineae). It mainly originates from Anqing and Bangbu of Anhui Province, Chongming and Qidong of Jiangsu Province, Hangzhou and Ningbo of Zhejiang Province, and Huanggang, Xiaogan and Jinzhou of Hubei Province.

According to studies on resources, *Rhizoma phragmitis* is planted abundantly in Baiyangdian District (Anxinxian, Xiongxian, etc) of Hebei Province.

Chemical Composition

Rhizoma phragmitis contains various chemical components, including amino acids, fatty acids, sterol, tocopherol, polyphenol, coixol, tricin, 5% proteins, 1% lipids, 51% carbohydrates, 0.1% asparagin, vitamin B_1, vitamin B_2 and vitamin C. The epidermal siliceous cells of *Rhizoma phragmitis* contain predominantly Si, followed by K, Cl, Ca, S, Fe, Cu, Zn and B.

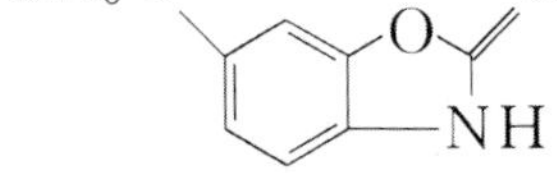

Coixol

Experimental Materials

Samples

Rhizoma phragmitis samples were collected from Baiyangdian District of Hebei Province in August 2002. There were 13 samples with variant identifications completed on all of them.

Reagents

Methanol (analytical-grade methanol was purchased from Shanghai Chemical Reagents Pte. Ltd.; chromatographic-grade methanol was purchased from Tedia Company Inc.). *n*-Butyl alcohol (analytical-grade, Shanghai Chemical Reagents Pte. Ltd.). Double-distilled deionized water (prepared by our lab). Phosphoric acid (analytical-grade).

Preparation of Samples

10 g of crude powder of the sample was accurately weighed and placed in a round-bottom flask. Eight times of water was added to the flask. After refluxing for 2 h, the extract was concentrated to 50 ml and was placed in a 125 ml separating funnel. 50 ml of *n*-butanol was added to the funnel. After being extracted with shaking, it was set aside for the layers to separate. The layer of *n*-butanol was collected and placed in an evaporating dish. It was evaporated to dryness over a 100°C water bath. The residue was dissolved in a proper amount of methanol and filtered through a 0.45 μm Millipore filter. After the volume was topped up to 10 ml, the sample was then ready to be used.

HPLC Conditions

Instrument: Agilent 1100 high-performance liquid chromatography and diode array detector
Mobile phase: Methanol and 0.01% phosphoric acid with two-step

gradient elution
Flow rate: 1.0 ml/min
Column: HYPERSIL C_{18} analytical column (250 mm × 4.6 mm, 10 μm)
Column temperature: 30°C
Detecting wavelength: 236 nm
Reference wavelength: 360 nm
Run time: 40 min
Injection volume: 10 μl

Methods

The sample solution was injected into the liquid chromatograph. The chromatogram was obtained under selected HPLC conditions. The experimental method was validated.

Mobile Phase Selection

A comparison of the experimental results showed that two-step gradient elution with the methanol and 0.01% phosphoric acid solution could successfully separate the chromatographic peaks and produce the maximum number of peaks from the samples. The duration of the analysis was 40 min. The 1.5 h chromatogram of the samples showed no characteristic peaks after 40 min.

A is methanol, B is 0.01% phosphoric acid.

	A (%)	B (%)
0–15 min	5–30	70–95
15–45 min	30–75	25–70
45–55 min	75–75	25–25

Method Validation

1. Stability: Test samples were stored at room temperature and tested at 0 h, 1 h, 2 h, 4 h, 6 h, 24 h and 48 h. The degrees of similarity

were analyzed by the Similarity Degree Calculation Software for Traditional Chinese Medicine Fingerprints. The results showed that the degrees of similarity were 0.9444–0.9987, which indicated that samples were stable within the 48 testing hours.

2. Precision: Five consecutive injections of the samples were conducted to determine the precision. The degrees of similarity were analyzed by the Similarity Degree Calculation Software for Traditional Chinese Medicine Fingerprints. The results showed that the degrees of similarity were 0.9799–0.9986, which indicated that the injections of the test samples were carried out with precision.
3. Reproducibility: Five samples from the same batch were prepared and tested according to the method described above. The degrees of similarity were analyzed by the Similarity Degree Calculation Software for Traditional Chinese Medicine Fingerprints. The results showed that the degrees of similarity were 0.9419–0.9986, indicating good reproducibility of the results.

Sample Analysis

The HPLC fingerprints of the test samples from 10 batches were analyzed under the conditions described above.

Establishment of Fingerprints

The number of peaks, peak area and relative retention time of 10 batches of samples were analyzed and compared to obtain an optimized fingerprint (Fig. 19.1).

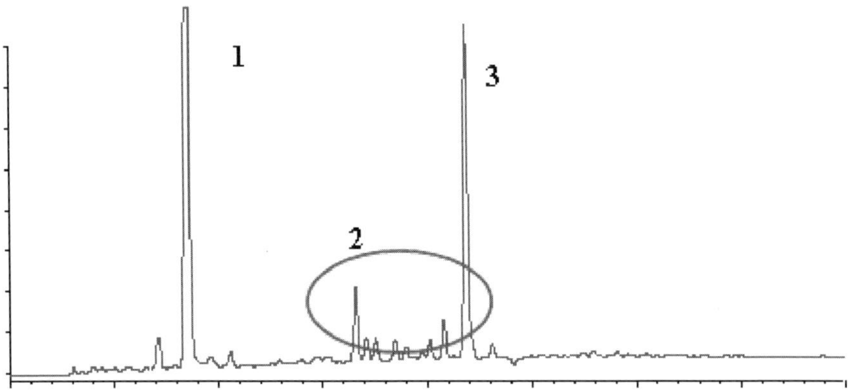

Fig. 19.1. HPLC fingerprint of Lugen (*Rhizoma phragmitis*).

Results and Discussion

More than 10 peaks were separated using HPLC. After comparing all the tested and documented fingerprints, three common peaks were selected as the characteristic peaks of *Rhizoma phragmitis*. The retention time (min) and peak area values of the three common peaks are listed below.

Peak number	1	2	3
Retention time	8.641	16.752	22.046
Peak area	83664.28	868.29	3245.70

Common Peak Standardization

Common peaks were assigned according to the results of 10 batches of samples. There were three common peaks in the HPLC fingerprints of *Rhizoma phragmitis*. Peak 1 of the test samples was used as the reference peak. Compared to the reference peak, the relative retention times of the other two common peaks were 1.939 and 2.551.

Ratio of Common Peak Area

Aside from the reference peak, the peak area of common peak 3 exceeded 10% of the total peak area. Its relative peak area ratio was specified as 0.0272–0.0504.

2.20

Lianqiao
连翘
Fructus forsythiae

Introduction

Fructus forsythiae is the dried fruit of *Forsythia suspensa* (Thunb.) Vahl (family Oleaceae). It mainly originates from Shanxi, Henan, Shaanxi and Shandong Provinces. Shanxi and Henan Provinces have the highest yield of *Fructus forsythiae* in China. There are two types of *Fructus forsythiae*. When the fruit is *en primeur* and still green, it is called Qingqiao (青翘). If the fruit is fully mature, it is called Laoqiao (老翘).

According to studies on resources, *Fructus forsythiae* is planted abundantly in Linbao, Yichuan, Qinyang, Huixian, Songxian, Mianchi, Shanxian, Lushi, Luoning, Jiyuan, Xiuwu, Xixia, Luanchuan, Nanzhao, Linxian and Zhechuan of Henan Province.

Chemical Composition

Fructus forsythiae contains forsythin, forsythigenin, forthiaside A, forsythol, rutin and caffeic acid.

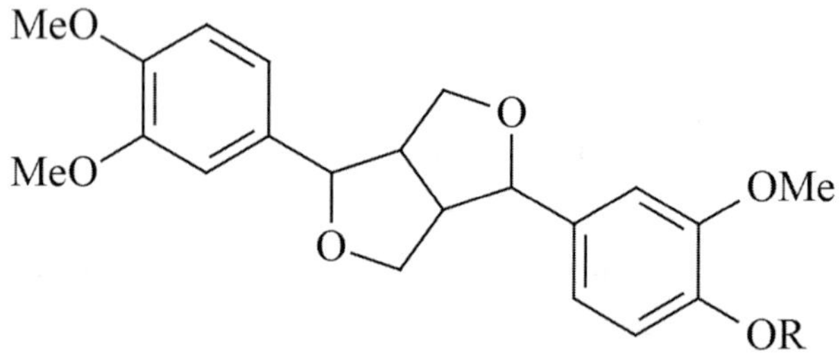

Forsythigenin R=H
Forsythin R=glu

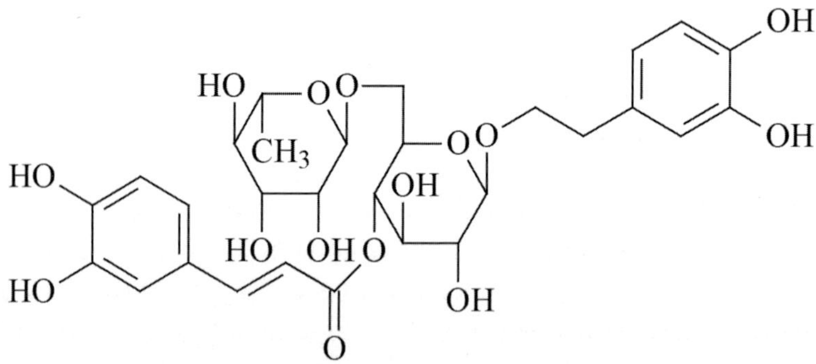

Forthiaside A

Experimental Materials

Samples

Green *Fructus forsythiae en primeur* samples were collected from four-year-old plants in Lingbao, Yichuan and Qinyang of Henan Province in early September 2001. There were 12 samples with variant identifications completed on all of them.

Reference

Forsythin for quantitative analysis was provided by the National Institute for the Control of Pharmaceutical and Biological Products (China).

Reagents

Methanol (analytical-grade methanol was purchased from Shanghai Chemical Reagents Pte. Ltd.; chromatographic-grade methanol was purchased from Tedia Company Inc.). Ether (analytical-grade, Shanghai Chemical Reagents Pte. Ltd.). Double-distilled deionized water (prepared by our lab). Phosphoric acid (analytical-grade).

Preparation of Samples

20 g of crude powder of the sample was accurately weighed and placed in a round-bottom flask. Ten times of water was added to the flask and the mixture was distilled for 80 min. The volatile oils, emulsion and extract (approximate 20 ml) were collected. The extract was concentrated to 50 ml and placed in a 125 ml separating funnel. 50 ml of ether was added to the funnel. After being extracted with shaking, it was left aside for the layers to separate. The ether layer was collected and was transferred to a 250 ml beaker. It was evaporated to dryness on a 65°C water bath. The residue was dissolved in a proper amount of methanol and filtered through a 0.45 μm Millipore filter. After being topped up to 10 ml, it was then ready to be used as a test sample.

Preparation of Reference

A proper amount of forsythin control was accurately weighed and placed in a 50 ml volumetric flask. It was made into a 300 μg/ml solution by adding methanol and was then ready to be used as a reference.

HPLC Conditions

Instument: Agilent 1100 high-performance liquid chromatography and diode array detector

Mobile phase: Methanol and 0.01% phosphoric acid with two-step gradient elution
Flow rate: 1.0 ml/min
Column: HYPERSIL C_{18} analytical column (250 mm × 4.6 mm, 10 μm)
Column temperature: 30°C
Detecting wavelength: 236 nm
Reference wavelength: 360 nm
Run time: 40 min
Injection volume: 20 μl sample, 10 μl reference

Methods

The sample solution was injected into the liquid chromatograph. The chromatogram was obtained under selected HPLC conditions. The experimental method was validated.

Mobile Phase Selection

A comparison of the experimental results showed that two-step gradient elution with methanol and 0.01% phosphoric acid solution could successfully separate the chromatographic peaks and produce the maximum number of peaks from the samples. The duration of the analysis was 40 min. The 1.5 h chromatogram of the samples showed no characteristic peaks after 40 min.

A is methanol, B is 0.01% phosphoric acid.

	A (%)	B (%)
0–15 min	5–30	70–95
15–45 min	30–75	25–70
45–55 min	75–75	25–25

Method Validation

1. Stability: Test samples were stored at room temperature and tested at 0 h, 1 h, 2 h, 4 h, 6 h, 24 h and 48 h. The degrees of similarity were analyzed by the Similarity Degree Calculation Software for Traditional Chinese Medicine Fingerprints. The results showed that the degrees of similarity were 0.9828–0.9987, which indicated that samples were stable within the 48 testing hours.
2. Precision: Five consecutive injections of the samples were conducted to determine the precision. The degrees of similarity were analyzed by the Similarity Degree Calculation Software for Traditional Chinese Medicine Fingerprints. The results showed that the degrees of similarity were 0.9897–0.9965, which indicated that the injections of the test samples were carried out with precision.
3. Reproducibility: Five samples from the same batch were prepared according to the method described above and were tested simultaneously. The degrees of similarity were analyzed by the Similarity Degree Calculation Software for Traditional Chinese Medicine Fingerprints. The results showed that the degrees of similarity were 0.9698–0.9993, indicating good reproducibility of the results.

Sample Analysis

The HPLC fingerprints of the test samples from 10 batches were analyzed under the same conditions described above.

Establishment of Fingerprints

The number of peaks, peak area and relative retention time of 10 batches of samples were analyzed and compared to obtain an optimized fingerprint (Fig. 20.1).

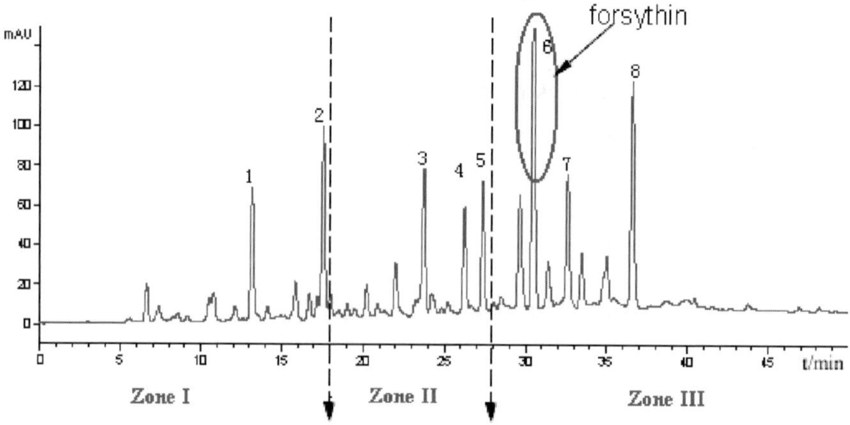

Fig. 20.1. HPLC fingerprint of Lianqiao (*Fructus forsythiae*).

Results and Discussion

More than 30 peaks were separated using HPLC. After comparing all the tested and documented fingerprints, eight common peaks were selected as the characteristic peaks of *Fructus forsythiae*. The retention time (min) and peak area values of the eight common peaks are listed below.

Peak number	1	2	3	4	5	6	7	8
Retention time	13.204	17.683	23.937	26.485	27.346	30.541	32.673	36.758
Peak area	1027.02	1395.10	1208.35	853.94	1047.14	2688.16	1472.85	1968.60

Common Peak Standardization

Common peaks were assigned according to the results of 10 batches of samples. There were eight common peaks in the HPLC fingerprints of *Fructus forsythiae*. Peak 1 of the test samples was used as the reference peak. Compared to the forsythin reference peak (peak 6), the relative retention times of the other seven common peaks were 0.432, 0.579, 0.784, 0.867, 0.895, 1.070 and 1.204.

Characteristics of Common Peak Area

The forsythin reference peak (peak 6) was the strongest peak. The whole chromatogram was divided into three zones. Zone I had two peaks, and peak 2 was stronger than peak 1. Zone II and zone III had three peaks each.

Ratio of Common Peak Area

Aside from the forsythin reference peak, no common peak area exceeded 10% of the total peak area.

2.21

Mudanpi
牡丹皮
Cortex moutan

Introduction

Cortex moutan is the dried root bark of *Paeonia suffruticosa* Andr. (family Ranunculaceae). It mainly originates from Anhui and Sichuan Provinces.

According to studies on resources, Anhui has the largest yield of *Cortex moutan* in China. Its *Cortex moutan* is sold all over China and is exported. It is planted abundantly in Tongling, Nanling, Qingyang, Fanchang, Tingxian of Anhui Province. The best-quality *Cortex moutan* is produced in Fenghuang Hill of Tongling in Anhui Province.

Chemical Composition

Cortex moutan mainly contains paeonol, paeonoside, paeonolide (paeonol-D-glucose-L-arabinose), paeoniflorin, oxypaeoniflorin, benzoylpaeoniflorin, and benzoyloxypaeonifloin. Some chemical structures are listed below.

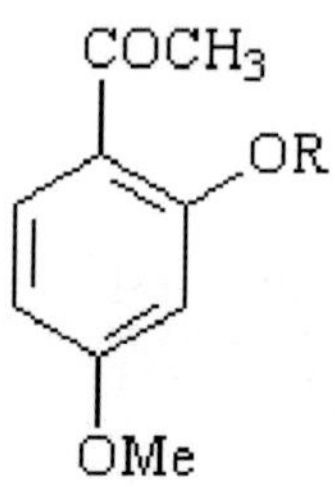

Paeonoside R=β-D-glu

Paeonolide R=β-D-glu-L-ara

Paeonol R=H

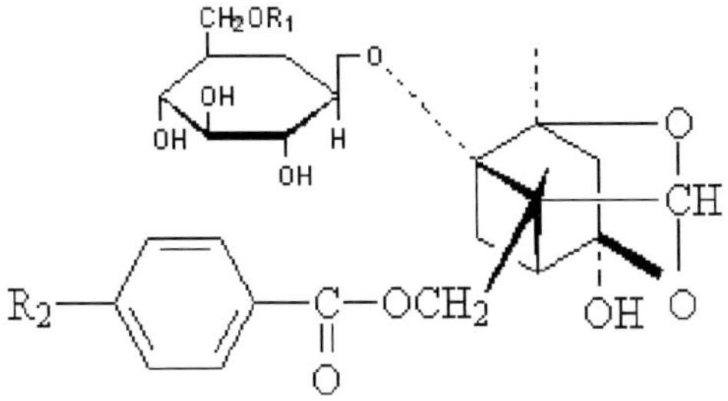

Paeoniflorin $R_1=R_2=H$

Oxypaeoniflorin $R_1=H$, $R_2=H$

Benzoylpaeoniflorin

$R_1=$ 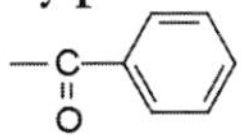 , $R_2=H$

Benzoyloxypaeonifloin

$R_1=$ 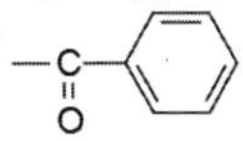 , $R_2=OH$

Experimental Materials

Samples

Cortex moutan samples were collected from Fenghuang Hill of Tongling and Ya Hill of Nanling in Anhui Province in late October 2001. There were 12 samples with variant identifications completed on all of them.

Reference

Paeonol for quantitative analysis was provided by the National Institute for the Control of Pharmaceutical and Biological Products (China).

Reagents

Acetonitrile (chromatographic-grade, Tedia Company Inc., USA). Methanol (analytical-grade methanol was purchased from Shanghai Chemical Reagents Pte. Ltd.; chromatographic-grade methanol was purchased from Tedia Company Inc.). Acetic acid (analytical-grade, Nanjing Chemical Reagent Co. Ltd.). Double-distilled deionized water (prepared by our lab).

Preparation of Samples

0.5 g of powder sample (passed through a 20 mesh sieve) was accurately weighed. After adding 20 ml of methanol, it was refluxed for 1 h and filtered. The residue was refluxed twice using the same method and filtered again. The filtrates were combined. The solvent was evaporated off at low heat. The residue was dissolved in methanol and topped up to 25 ml. The solution was mixed well and filtered through a 0.45 μm Millipore filter. After discarding the first 5 ml filtrate, the remaining filtrate was ready to be used as a test sample.

Preparation of Reference

Paeonol was accurately weighed and dissolved in methanol to make it into a 20 μg/ml solution. It was then ready to be used as a reference.

HPLC Conditions

Instrument: Agilent 1100 high-performance liquid chromatography and diode array detector
Column: Inertsil® ODS-3 analytical column (250 mm × 4.6 mm, 5 μm)
Mobile phase: Gradient elution with 1% HAc-H_2O and 1% HAc-CH_3CN
Flow rate: 1.0 ml/min
Column temperature: 30°C
Detecting wavelength: 256 nm
Reference wavelength: 360 nm
Run time: 60 min
Injection volume: 10 μl sample, 10 μl reference

Methods

Proper amount of sample solution was injected into the liquid chromatograph. The chromatogram was obtained under selected HPLC conditions. The experimental method was validated.

Mobile Phase Selection

A comparison of the experimental results showed that gradient elution with 1% HAc-H_2O and 1% HAc-CH_3CN could successfully separate the chromatographic peaks and produce the maximum number of peaks from the samples. The duration of the analysis was 60 min. The 1.5 h chromatogram of the samples showed no characteristic peaks after 60 min.

A is 1% HAc-H_2O, B is 1% HAc-CH_3CN.

	A (%)	B (%)
0–3 min	95–95	5–5
3–35 min	70–95	5–30
35–55 min	40–70	30–60
55–60 min	40–40	60–60

Method Validation

1. Stability: Test samples from the same batch were stored at room temperature and tested at 0 h, 2 h, 4 h, 7 h, 11 h, 22 h, 29 h and 36 h. The degrees of similarity calculated by the Similarity Degree Calculation Software for Traditional Chinese Medicine Fingerprints (DEMO version) were all above 0.95, which indicated that samples were stable within the 36 testing hours.
2. Precision: Five consecutive injections of the samples (from the same batch) were conducted to determine the precision. The degrees of similarity calculated by the Similarity Degree Calculation Software for Traditional Chinese Medicine Fingerprints (DEMO version) were all above 0.99, which indicated that the injections of the test samples were carried out with precision.
3. Reproducibility: Five samples from the same batch were prepared according to the method described above and were tested simultaneously. The degrees of similarity calculated by the Similarity Degree Calculation Software for Traditional Chinese Medicine

Fingerprints (DEMO version) were all above 0.95, indicating good reproducibility of the results.

Sample Analysis

The HPLC fingerprints of the test samples from 10 batches were analyzed under the conditions described above.

Establishment of Fingerprints

The number of peaks, peak area and relative retention time of 10 batches of samples were analyzed and compared to obtain an optimized fingerprint (Fig. 21.1).

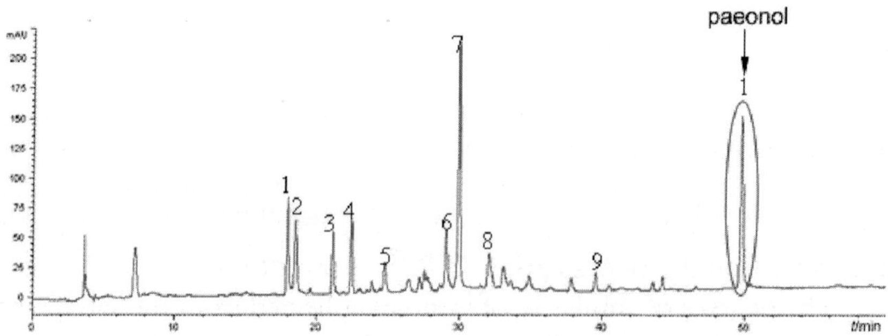

Fig. 21.1. HPLC fingerprint of Mudanpi (*Cortex moutan*).

Results and Discussion

About 26 peaks were separated using HPLC. After comparing all the tested and documented fingerprints, 10 common peaks were selected as the characteristic peaks of *Cortex moutan*. The retention time (min) and peak area values of the 10 common peaks are listed below.

Peak number	1	2	3	4	5	6	7	8	9	10
Retention time	17.88	18.45	21.03	22.32	24.68	28.94	29.82	31.96	39.44	49.86
Peak area	879.81	873.34	545.71	705.64	371.76	641.44	2278.67	486.8	178.46	1775.75

Common Peak Standardization

Common peaks were assigned according to the results of 10 batches of samples. There were 10 common peaks in the HPLC fingerprints of *Cortex moutan*. Compared to the paeonol reference peak (peak 10), the relative retention times of the other nine common peaks were 0.359, 0.370, 0.422, 0.448, 0.495, 0.580, 0.598, 0.641 and 0.791.

Ratio of Common Peak Area

Aside from the paeonol reference peak, the peak area of common peak 7 was more than 10% of the total peak area. The relative peak area ratio of peak 7 was specified as 0.898–1.668.

Evaluation of Similarity

Experimental data were imported into the TCM Fingerprint Identification and Evaluation System. After selecting peaks and setting a matching template, the chromatogram peaks were automatically matched. Then the standard template was determined, and was used for overall similarity evaluation of fingerprint peaks.

The degrees of similarity of samples collected from the same place (Fenghuang Hill in Tongling of Anhui Province) but different growth phases were compared. The results showed that three-year-old and four-year-old *Cortex moutan* were nearly the same. And they were only slightly different from the one-year-old fresh *Cortex moutan*. The concentrations of the components in one-year-old fresh *Cortex moutan* are obviously less than those in three-year-old and four-year-old ones. In some places, the local people called the one-year-old *Cortex moutan* seedlings, which were normally not used as a medicinal herb. Those used for medicinal purposes were at least four-year-old or five-year-old plants.

The similarity evaluation results also showed that there were big differences between fresh and dried (sample 2 Guadanpi is also a dried sample) *Cortex moutan*. Fresh *Cortex moutan* contained more volatile components than dried *Cortex moutan*, which could be identified by the smell of the samples. The aroma of fresh *Cortex moutan* was stronger than that of dried *Cortex moutan*. Analysis showed that there were more white crystal precipitations on the inner surface of the fresh root bark.

2.22

Heshouwu

何首乌

Radix Polygoni multiflori

Introduction

Radix Polygoni multiflori is the dried root tuber of *Polygonum multiflorum* Thunb. (family Polygonaceae). It mainly originates from Henan, Hubei, Guangxi, Guangdong, Guizhou, Sichuan and Jiangsu. It also originates from Hunan, Shaanxi, Zhejiang, Anhui, Jiangxi, Shandong and Yunnan.

According to some studies, *Radix Polygoni multiflori* is planted abundantly in Liu'an and Quanjiao of Anhui Province.

Chemical Composition

The root tuber contains anthraquinones, which mainly are chrysophanol, emodin, rhein, physcion and chrysophanolanthrone. In addition, it contains stilbene glucoside, such as 2,3,5,4′-tetrahydroxystilbene-2-*O*-*β*-D-glucoside.

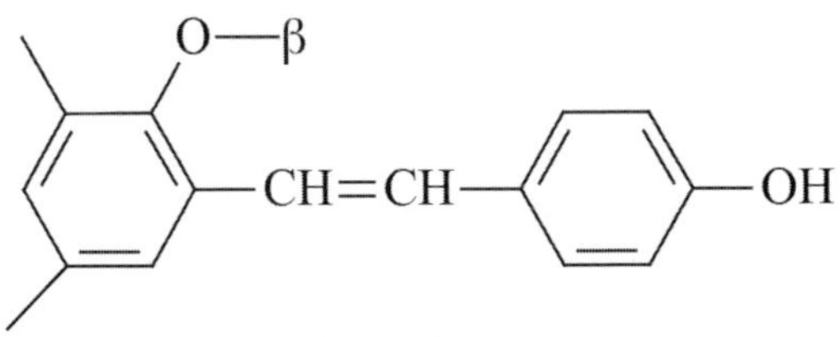

2,3,5,4′-Tetrahydroxystilbene-2-*O*-*β*-D-glucoside

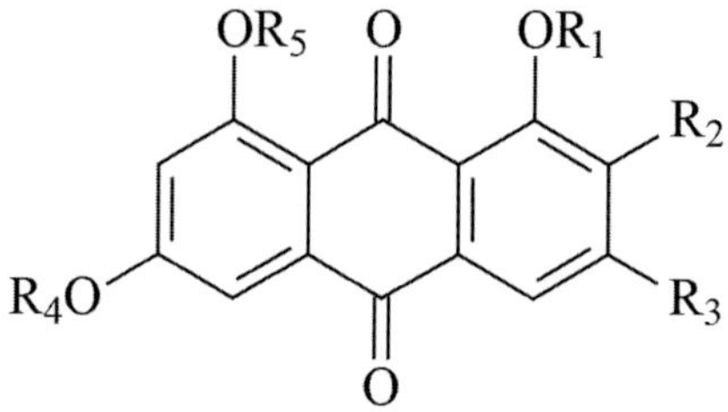

Emodin $R_1=R_2=R_4=R_5=H$, $R_3=CH_3$
Physcion $R_1=R_2=R_5=H$, $R_3=R_4=CH_3$
Emodin-1,6-dimethyl-ether $R_1=R_3=R_4=CH_3$, $R_2=R_5=H$
Emodin-8-methyl-ether $R_1=R_2=R_4=H$, $R_3=R_5=CH_3$
6-Hydroxyl-emodin $R_1=R_2=R_4=R_5=H$, $R_3=CH_2OH$
6-Hydroxyl-emodin-8-methyl-ether $R_1=R_2=R_4=H$, $R_3=CH_2OH$, $R_5=CH_3$
2-Acetyl-emodin $R_1=R_4=H$, $R_2=CH_2OH$, $R_3=R_5=CH_3$
Emodin-8-*O*-*β*-D-glucoside $R_1=R_4=H$, $R_2=CH_2OH$, $R_3=CH_3$, $R_5=glu$
Physcion-8-*O*-*β*-D-glucoside $R_1=H$, $R_2=CH_2OH$, $R_3=R_4=CH_3$, $R_5=glu$

Experimental Materials

Samples

12 samples of cultivated species were collected from Liu'an District of Anhui province. Samples of wild species were collected from Laoshan in Jiangpu of Nanjing. Variant identifications were completed on all of them.

Reference

Emodin was prepared by our lab.

Reagents

Acetonitrile (chromatographic-grade, Tedia Company Inc., USA). Double-distilled deionized water (prepared by our lab). Ethanol (analytical-grade ethanol was purchased from Shanghai Chemical Reagents Pte. Ltd.; chromatographic-grade ethanol was purchased from Tedia Company Inc.). Chloroform (analytical-grade, Shanghai Chemical Reagents Pte. Ltd.). Isopropanol (analytical-grade, Shanghai Chemical Reagents Pte. Ltd.).

Preparation of Samples

5 g of crude powder of the sample was accurately weighed and placed in a round-bottom flask. 100 ml of ethanol was added to the flask. After refluxing three times, the filtrates were combined and concentrated to a small volume. It was then extracted twice with 40 ml of chloroform. The extracts were combined and dehydrated by anhydrous sodium sulfate. After being filtered and concentrated to a proper volume, it was evaporated to dryness. The residue was dissolved in isopropanol and topped up to 25 ml in a volumetric flask. After being filtered through a 0.45 μm Millipore filter, the filtrate was then ready to be used as a test sample.

Preparation of Reference

Emodin was accurately weighed and dissolved in methanol to make into a 20 μg/ml solution. It was then ready to be used as a reference.

HPLC Conditions

Instrument: Agilent 1100 high-performance liquid chromatography and diode array detector
Mobile phase: Gradient elution with water and acetonitrile
Flow rate: 1.0 ml/min
Column: Agilent ZORBAX 80A Extend-C_{18} analytical column (250 mm × 4.6 mm, 5 μm)
Column temperature: 25°C
Detecting wavelength: 230 nm
Reference wavelength: 360 nm
Run time: 80 min
Injection volume: 10 μl sample, 10 μl reference

Methods

A proper amount of sample solution was injected into the liquid chromatograph. The chromatogram was obtained under selected HPLC conditions. The experimental method was validated.

Mobile Phase Selection

A comparison of the experimental results showed that gradient elution with water and acetonitrile could successfully separate the chromatographic peaks and produce the maximum number of peaks from the samples. The duration of the analysis was 80 min. The 2 h chromatogram of the samples showed no characteristic peaks after 80 min.

A is water, B is acetonitrile.

	A (%)	B (%)
0–60 min	0–98	2–100
60–80 min	0–0	100–100

Method Validation

1. Stability: Test samples from the same batch were stored at room temperature and injected at regular intervals for testing. The degrees of similarity calculated by the Similarity Degree Calculation Software for Traditional Chinese Medicine Fingerprints were all above 0.91, which indicated that samples were stable within the 36 testing hours.
2. Precision: Five consecutive injections of the samples from the same batch were carried out to determine the precision. The degrees of similarity calculated by the Similarity Degree Calculation Software for Traditional Chinese Medicine Fingerprints were all above 0.96, which indicated that the injections of the test samples were carried out with precision.
3. Reproducibility: Five samples from the same batch were prepared and tested according to the method described above. The degrees of similarity calculated by the Similarity Degree Calculation Software for Traditional Chinese Medicine Fingerprints were all above 0.92, indicating good reproducibility of the results.

Sample Analysis

The HPLC fingerprints of the test samples from 10 batches were analyzed under the conditions described above.

Establishment of Fingerprints

The number of peaks, peak area and relative retention time of 10 batches of samples were analyzed and compared to obtain an optimized fingerprint (Figs. 22.1, 22.2).

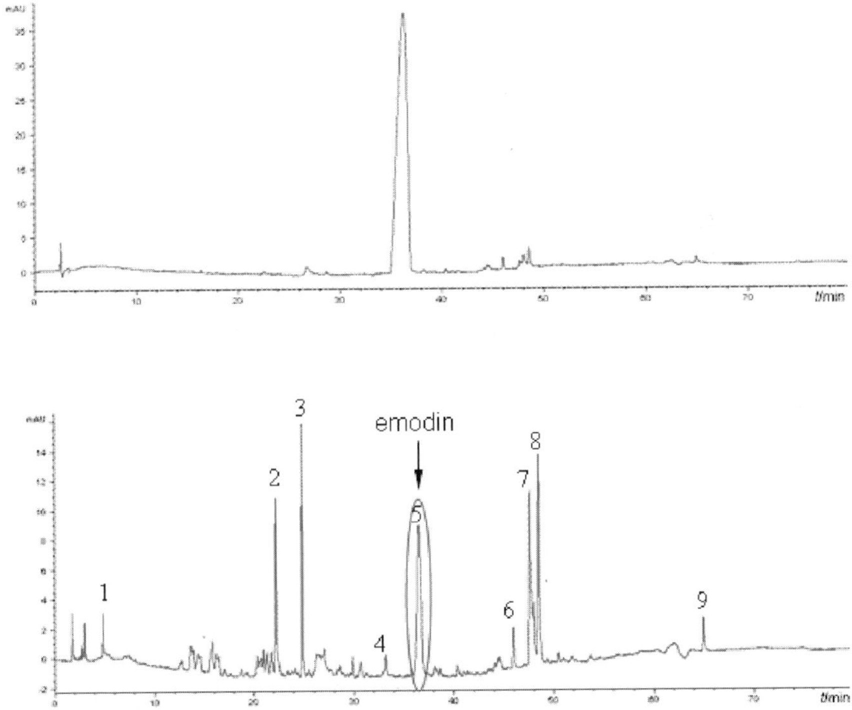

Fig. 22.1. HPLC fingerprints of emodin reference (top) and of Heshouwu (*Radix Polygoni multiflori*) (bottom).

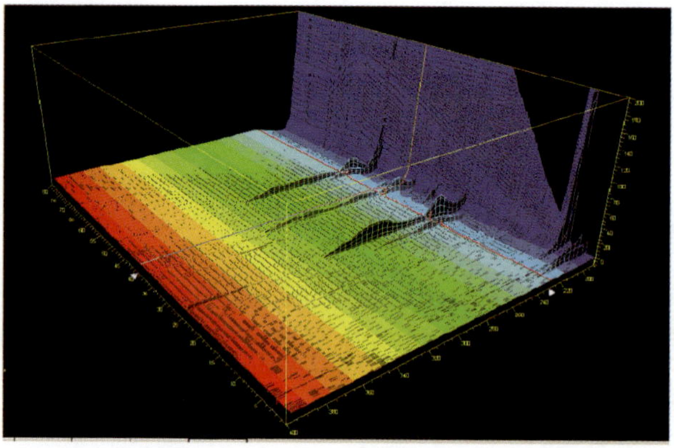

Fig. 22.2. 3-D HPLC fingerprint of Heshouwu (*Radix Polygoni multiflori*).

Results and Discussion

Over 30 peaks were separated using HPLC. After comparing with the reference and all the tested and documented fingerprints, nine common peaks were selected as the characteristic peaks of the *Radix Polygoni multiflori* fingerprint. The retention time (min) and peak area values of the nine common peaks are listed below.

Peak number	1	2	3	4	5	6	7	8	9
Retention time	4.787	22.169	24.773	33.08	36.434	45.936	47.579	48.482	64.964
Peak area	11.2	109.4	119.9	14.4	269.6	27.6	153.3	167.7	21.7

Common Peak Standardization

Common peaks were assigned according to the results of 10 batches of samples. There were nine common peaks in the HPLC fingerprints of *Radix Polygoni multiflori*. Compared to the emodin reference peak (peak 5), the relative retention times of the other eight common peaks were 0.131, 0.608, 0.680, 0.908, 1.261, 1.306, 1.331 and 1.783.

Ratio of Common Peak Areas

Aside from the emodin reference peak, the peak areas of common peaks 3, 7 and 8 exceeded 10% of the total peak area. The relative peak area ratio of peak 3 is specified as 0.312–0.579. The relative peak area ratio of peak 7 is specified as 0.398–0.740. The relative peak area ratio of peak 8 is specified as 0.435–0.809.

2.23

Zhiheshouwu
制何首乌
Radix Polygoni multiflori Preparata

Introduction

Radix Polygoni multiflori Preparata is the processed dried root tuber of *Polygonum multiflorum* Thunb. (family Polygonaceae).

Origination

q.v. "Heshouwu (*Radix Polygoni multiflori*)".

Chemical Composition

q.v. "Heshouwu (*Radix Polygoni multiflori*)".

Experimental Materials

Samples

Samples were bought from Liu'an District of Anhui Province. There were 10 samples with variant identifications completed on all of them.

Reagents

Acetonitrile (chromatographic-grade, Tedia Company Inc., USA). Double-distilled deionized water (prepared by our lab). Ethanol (analytical-grade ethanol was purchased from Shanghai Chemical Reagents Pte. Ltd.; chromatographic-grade ethanol was purchased from Tedia Company Inc.). Chloroform (analytical-grade, Shanghai Chemical Reagents Pte. Ltd.). Isopropanol (analytical-grade, Shanghai Chemical Reagents Pte. Ltd.).

Preparation of Samples

5 g of crude powder of the sample was accurately weighed and placed in a round-bottom flask. 100 ml of ethanol was added to the flask. After refluxing three times, the filtrates were combined and concentrated to a minimum volume. It was then extracted twice with 40 ml of chloroform. The extracts were combined and dehydrated by anhydrous sodium sulfate. After being filtered and concentrated to a proper volume, it was evaporated to dryness. The residue was dissolved in isopropanol and was topped up to 25 ml in a volumetric flask. After being filtered through a 0.45 μm Millipore filter, the filtrate was then ready to be used as a test sample.

HPLC Conditions

Instrument: Agilent 1100 high-performance liquid chromatography and diode array detector
Mobile phase: Gradient elution with water and acetonitrile
Flow rate: 1.0 ml/min
Column: Agilent ZORBAX 80A Extend-C_{18} analytical column (250 mm × 4.6 mm, 5 μm)
Column temperature: 25°C
Detecting wavelength: 210 nm
Reference wavelength: 360 nm
Run time: 80 min
Injection volume: 10 μl

Methods

A proper amount of sample solution was injected into the liquid chromatograph. The chromatogram was obtained under selected HPLC conditions. The experimental method was validated.

Mobile Phase Selection

A comparison of the experimental results showed that gradient elution with water and acetonitrile could successfully separate the chromatographic peaks and produce the maximum number of peaks from the samples. The duration of the analysis was 80 min. The 2 h chromatogram of the samples showed no characteristic peaks after 80 min.

A is water, B is acetonitrile.

	A (%)	B (%)
0–70 min	0–95	5–100
70–80 min	0–0	100–100

Method Validation

1. Stability: Test samples from the same batch were stored at room temperature and injected at regular intervals for testing. The degrees of similarity calculated by the Similarity Degree Calculation Software for Traditional Chinese Medicine Fingerprints were all above 0.93, which indicated that samples were stable within the 36 testing hours.
2. Precision: Five consecutive injections of the samples from the same batch were carried out to determine the precision. The degrees of similarity calculated by the Similarity Degree Calculation Software for Traditional Chinese Medicine Fingerprints were all above 0.95, which indicated that the injections of the test samples were carried out with precision.
3. Reproducibility: Five samples from the same batch were prepared and tested according to the method described above. The degrees of similarity calculated by the Similarity Degree Calculation Software for Traditional Chinese Medicine Fingerprints were all above 0.91, indicating good reproducibility of the results.

Sample Analysis

The HPLC fingerprints of the test samples from 10 batches were analyzed under the conditions described above.

Establishment of Fingerprints

The number of peaks, peak area and relative retention time of 10 batches of samples were analyzed and compared to obtain an optimized fingerprint (Figs. 23.1, 23.2).

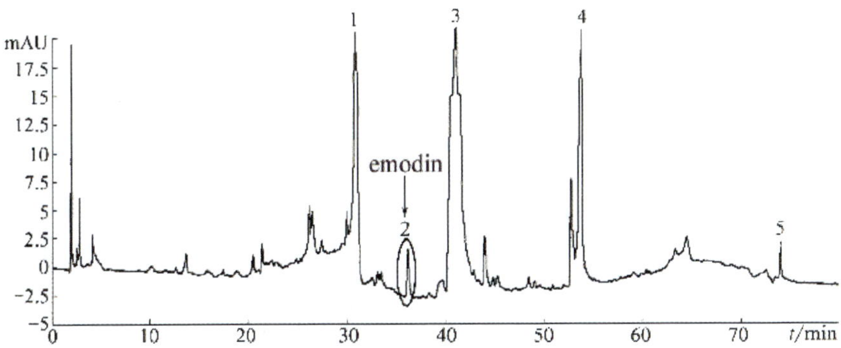

Fig. 23.1. HPLC fingerprint of Zhiheshouwu (*Radix Polygoni multiflori Preparata*).

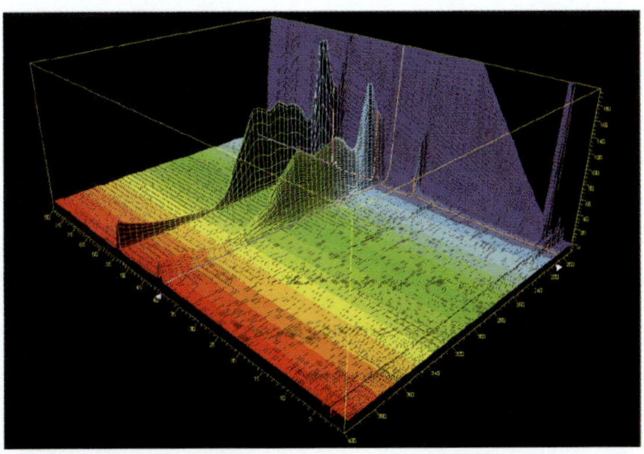

Fig. 23.2. 3-D HPLC fingerprint of Zhiheshouwu (*Radix Polygoni multiflori Preparata*).

Results and Discussion

About 20 peaks were separated using HPLC. After comparing with reference controls and comparing all the tested and documented fingerprints, five common peaks were selected as the characteristic peaks of the *Radix Polygoni multiflori Preparata* fingerprint. The retention time (min) and peak area values of the five common peaks are listed below.

Peak number	1	2	3	4	5
Retention time	30.831	36.141	41.148	53.678	73.958
Peak area	909.380	73.944	4125.628	1480.003	50.354

Common Peak Standardization

Common peaks were assigned according to the results of 10 batches of samples. There were five common peaks in the HPLC fingerprints of *Radix Polygoni multiflori Preparata*. Compared to the emodin reference peak (peak 2), the relative retention times of the other four common peaks were 0.853, 1.138, 1.485 and 2.041.

Ratio of Common Peak Areas

Aside from the reference peak, the peak areas of common peaks 1, 3 and 4 exceeded 10% of the total peak area. The relative peak area ratio of peak 1 is specified as 0.597–1.109. The relative peak area ratio of peak 3 is specified as 0.797–1.479. The relative peak area ratio of peak 4 is specified as 1.040–1.9305.

2.24

Buguzhi
补骨脂
Fructus psoraleae

Introduction

Fructus psoraleae is the dried mature fruit of *Psoralea corylifolia* L. (family Leguminosae). It mainly originates from Sichuan and Henan Provinces. It is called Chuanguzi (川故子) if it originates from Sichuan, and Huaiguzi (怀故子) if it originates from Henan.

According to studies on resources, *Fructus psoraleae* is planted abundantly in Jintang, Guangyuan and Dujiangyan (Guanxian) of Sichuan, and Hechuan and Jiangjin of Chongqing. Most of the commercially available *Fructus psoraleae* is from these places.

Chemical Composition

Fructus psoraleae contains volatile oils, coumarins, flavones, monoterpene phenols, lipid compounds, resin and stigmasterol. Their main components are listed below.

1. Volatile oils: limonene, terpineol-4, linalool, β-caryophyllene and geranyl acetate.
2. Coumarins: psoralen, isopsoralen, psoralidin, isopsoralidin, corylidin and coumarone.
3. Flavones: corylifolin, corylifolinin, bavachinin, isobavachin, isobavachalcone, neobavaisoflavone, bavachromene, corylin, bavachalcone and neobavachalcone.

4. Monoterpenphenol: bakuchiol.
5. Lipid compounds: triglyceride, diglyceride and monoester.
6. Others: it also contains free fatty acids, stigmasterol, daucosterol, triacontane, glucose and saponin.

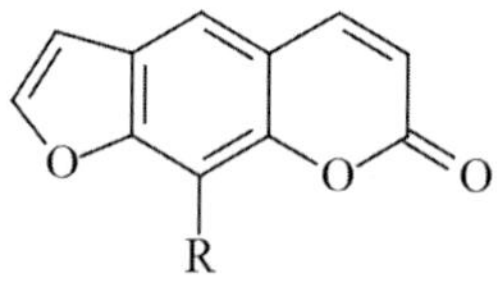

Psoralen

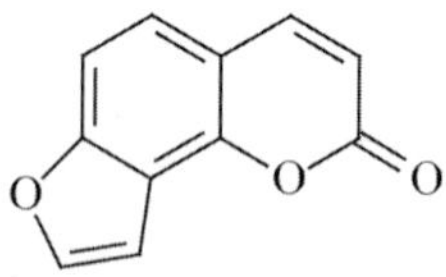

Isopsoralen

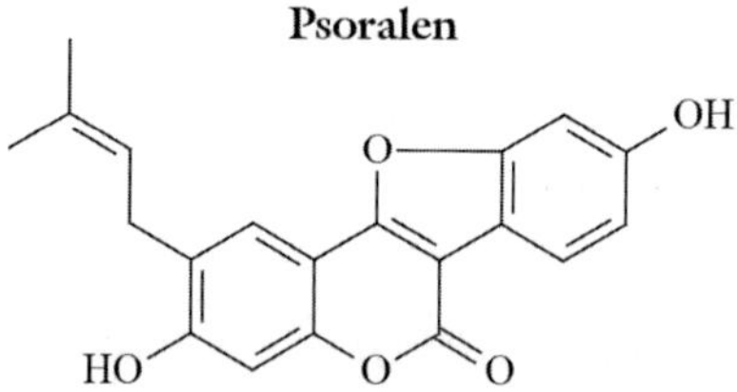

Psoralidin

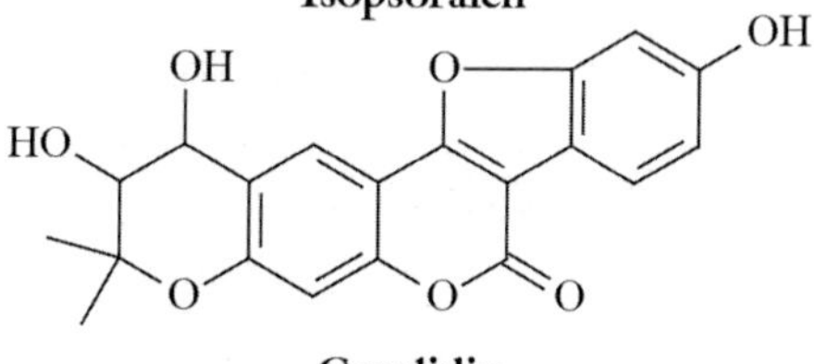

Corylidin

Experimental Materials

Samples

There were 13 samples collected from Jiangxi; Shanxi; Jiangjin and Jintang of Sichuan; Xinxiang, Shangqiu, Bo'ai and Xinyang of Henan; Anhui; Hunan; Hechuan of Sichuan; Yunnan; and Guangxi. Variant identifications were completed on all of them.

References

Psoralen and isopsoralen controls were provided by the National Institute for the Control of Pharmaceutical and Biological Products (China). Bavaisoflavone was prepared by our lab and the purity was 98%.

Reagents

Methanol and acetonitrile were chromatographic-grade. Other reagents were analytical-grade.

Preparation of Samples

5 g of crude powder of the sample was accurately weighed and bathed in a round-bottom flask for 1 h. After adding 80 ml of 55% ethanol, it was refluxed for 45 min. The sample solution was then transferred to a 250 ml volumetric flask and topped up to volume with distilled water. The solution was filtered and the filtrate was then ready to be used as a test sample.

Preparation of References

Psoralen was accurately weighed and dissolved in methanol to make a 0.1 mg/ml solution. Isopsoralen was accurately weighed and dissolved in methanol to make a 0.13 mg/ml solution. Bavaisoflavone was accurately weighed and dissolved in methanol to make a 0.14 mg/ml solution.

HPLC Conditions

Instrument: Jasco PU 980 high-performance liquid chromatography, Jasco PU 975 ultraviolet-visible detector (JASCO Corporation, Japan); CK chrom workstation (Scientific System Inc.)
Mobile phase: Gradient elution with 30 mmol/L methanol and acetate buffer (pH = 4.0)
Flow rate: 1.0 ml/min
Column: Hypersil ODS (150 mm × 4.6 mm, 5 μm), Dalian Elite Scientific Instruments Co., Ltd.
Column temperature: 25°C
Detecting wavelength: 320 nm
Run time: 50 min
Injection volume: 20 μl sample, 20 μl reference

Methods

A proper amount of sample solution was injected into the liquid chromatograph. The chromatogram was obtained under selected HPLC conditions. The experimental method was validated.

Mobile Phase Selection

A comparison of the experimental results showed that gradient elution with 30 mmol/L methanol and acetate buffer (pH = 4.0) could successfully separate the chromatographic peaks and produce the maximum number of peaks from the samples. The duration of the analysis was 50 min. The 1.5 h chromatogram of the samples showed no characteristic peaks after 50 min.

A is methanol (30 mmol/L), B is acetate buffer (pH = 4.0).

	A (%)	B (%)
0–5 min	25–25	75–75
5–50 min	25–95	5–75

Method Validation

1. Stability: *Fructus psoraleae* samples collected from Hechuan of Sichuan Province were tested at 0 h, 2 h, 4 h, 8 h, 12 h and 24 h. The RSD of the relative retention times of the peaks was less than 3.2% and the RSD of the relative peak areas was less than 3.7%, which indicated that samples were stable within the 24 testing hours.
2. Precision: *Fructus psoraleae* samples collected from Hechuan of Sichuan Province were used as test samples. Five consecutive injections of the samples were conducted to determine the precision. The RSD of the relative retention times of the peaks was less than 1.4% and the RSD of the relative peak areas was less than 2.0%, which indicated that the injections of the test samples were carried out with precision.
3. Reproducibility: Fingerprints of five samples from Hechuan of Sichuan Province, five samples from Xinxiang of Henan Province, and five samples from Jintang of Sichuan Province were established. The RSD of the relative retention times of the peaks was less than 0.72% and the RSD of the relative peak areas was less than 4.9%, indicating good reproducibility of the results.

Sample Analysis

The HPLC fingerprints of the test samples from 10 batches were analyzed under the conditions described above.

Establishment of Fingerprints

The number of peaks, peak area and relative retention time of 10 batches of test samples were analyzed and compared to obtain an optimized fingerprint (Fig. 24.1).

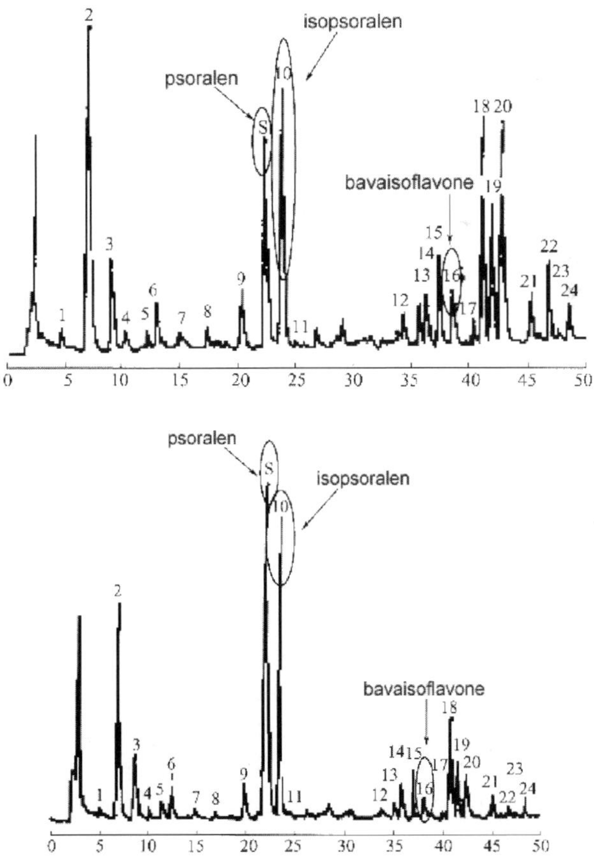

Fig. 24.1. HPLC fingerprints of Buguzhi (*Fructus psoraleae*) from Hechuan of Sichuan Province (top) and Bo'ai of Henan Province (bottom).

Results and Discussion

Over 40 peaks were separated using HPLC. After comparing with the reference peak and comparing all the tested and documented fingerprints, 25 common peaks were selected as the characteristic peaks of *Fructus psoraleae*.

Common Peak Standardization

Common peaks were assigned according to the results of 10 batches of samples. There were 25 common peaks in the HPLC fingerprints of *Fructus psoraleae*. Compared to the psoralen reference peak (peak S), the relative retention times of the other 24 common peaks were 0.193, 0.316, 0.405, 0.468, 0.545, 0.582, 0.678, 0.775, 0.915, 1.060, 1.119, 1.533, 1.592, 1.623, 1.670, 1.722, 1.808, 1.839, 1.878, 1.910, 2.032, 2.086, 2.125 and 2.175.

Ratio of Common Peak Areas

Aside from the psoralen reference peak, no other common peak had an area that exceeded 10% of the total peak area.

Evaluation of Similarity

The average value of *Fructus psoraleae* samples collected from 13 different places was used as a reference template. According to the peak matching results, the peak areas of the test samples and reference template were used to calculate the overall degrees of similarity. The degrees of similarity of *Fructus psoraleae* samples from 13 different places are as follows: Hunan 1.00, Bo'ai of Henan 0.99, Xinyang of Henan 0.99, Yunnan 0.96, Guangxi 0.93, Shangqiu of Henan 0.90, Jiangxi 0.90, Hechuan of Sichuan 0.88, Jiangjin of Sichuan 0.87, Anhui 0.87, Xinxiang of Henan 0.85, Jintang of Sichuan 0.83, Shanxi 0.75.

The fingerprint patterns of the main peaks of *Fructus psoraleae* samples from 13 different places were nearly the same. But there were some differences among the relative ratios of the concentrations of different components. The results showed that the degrees of similarity of samples from seven places were above 0.9, while those from the other six places were less than 0.9. A comparison of the fingerprints of *Fructus psoraleae* from Hunan (degree of similarity of 1.0) and Jintang of Sichuan (degree of similarity of 0.83) showed that the peak areas of psoralen (peak S) and isopsoralen (peak 10) in the Hunan sample were significantly smaller than those in the Jintang sample. The peak areas of peaks 19, 20 and 21 in the Hunan sample were significantly bigger than those in the Jintang sample. These resulted in the discrepancy in degree of similarity between the two samples.

2.25

Chenpi
陈皮
Pericarpium citri reticulatae

Introduction

Pericarpium citri reticulatae is the dried peel of the mature fruit of *Citrus reticulata* Blanco (family Rutaceae) and its cultivated variant. There are two types: one is Chenpi (陈皮), the other is Guangchenpi (广陈皮). Chenpi is also named Jupi (橘皮, tangerine peel) and originates from all tangerine-planting areas in China. Guangchenpi mainly originates from Xinhui, Jiangmen and Sihui of Guangdong Province.

Chemical Composition

Pericarpium citri reticulatae mainly contains volatile oils, flavones and other components. Flavones include hesperidin, neohesperidin, nobiletin, naringin and narirutin. *Pericarpium citri reticulatae* also contains synephrine. Some chemical structures are listed below.

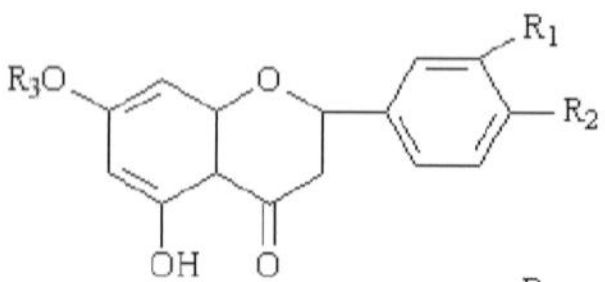

	R_1	R_2	R_3
Hesperidin	**OH**	**OCH_3**	**glu$^{6-1}$ rha**
Narirutin	**H**	**OH**	**glu$^{6-1}$ rha**

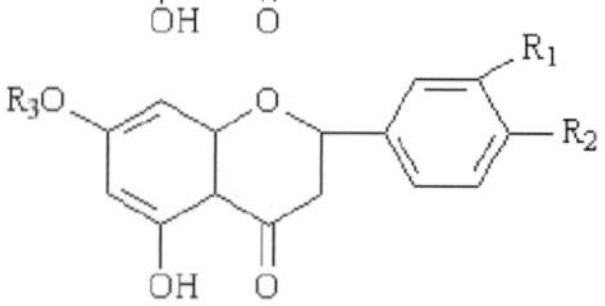

	R_1	R_2	R_3
Neohesperidin	**OH**	**OCH_3**	**glu$^{2-1}$ rha**
Naringin	**H**	**OH**	**glu$^{2-1}$ rha**

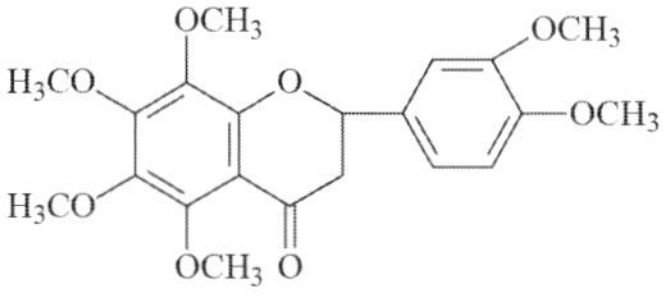

Nobiletin

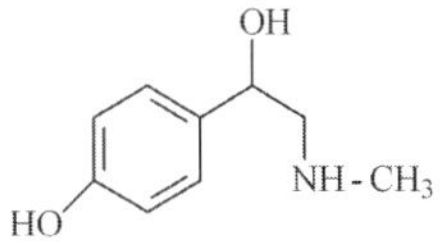

Synephrine

Experimental Materials

Samples

Guangchenpi samples were collected from Xinhui and Jiangmen of Guangdong Province in November 2002. There were 12 samples with variant identifications completed on all of them.

Reference

Synephrine used for quantitative analysis was provided by the National Institute for the Control of Pharmaceutical and Biological Products (China).

Reagents

Acetonitrile (chromatographic-grade, Tedia Company Inc., USA). Methanol (analytical-grade methanol was purchased from Shanghai Chemical Reagents Pte. Ltd.; chromatographic-grade methanol was purchased from Tedia Company Inc.). Phosphoric acid (HPLC-grade, Tedia Company Inc., USA). Double-distilled deionized water (prepared by our lab).

Preparation of Samples

0.5 g of powder sample was accurately weighed. After adding 15 ml of methanol, it was sonicated (operating frequency 40 kHz, ultrasound power 500 W) for 30 min. It was then filtered and the extraction was repeated twice. The filtrates were combined and topped up to 50 ml. The solution was mixed well and filtered again. The initial filtrate was discarded. A 2 ml aliquot of the remaining filtrate was collected and topped up to 5 ml with methanol in a volumetric flask. The solution was mixed well and ready to be used as a test sample.

Preparation of Reference

1 mg of synephrine was accurately weighed, dissolved and topped up to 100 ml with methanol in a volumetric flask. After being mixed well, it was ready to be used as a reference.

HPLC Conditions

Instrument: Agilent 1100 high-performance liquid chromatography and variable wave detector
Column: ZORBAX SB-C_{18} (4.6 mm × 250 mm, 5 μm)
Mobile phase: Gradient elution with methanol and aqueous solution of 0.02% H_3PO_4 and 0.1% sodium lauryl sulfate
Flow rate: 0.6 ml/min
Column temperature: 30°C
Detecting wavelength: 224 nm
Run time: 60 min
Injection volume: 20 μl sample, 20 μl reference

Methods

The sample solution was injected into the liquid chromatograph. The chromatogram was obtained under selected HPLC conditions. The experimental method was validated.

Mobile Phase Selection

A comparison of the experimental results showed that when methanol and an aqueous solution of 0.02% H_3PO_4 and 0.1% sodium lauryl sulfate were used as the mobile phase, gradient elution could successfully separate the chromatographic peaks and produce the maximum number of peaks. The duration of the analysis was 60 min. The 1.5 h chromatogram of the samples showed no characteristic peaks after 60 min.

A is methanol, B is aqueous solution of 0.02% H_3PO_4 and 0.1% sodium lauryl sulfate.

	A (%)	B (%)
0–25 min	5	95
25–35 min	55	45
35–60 min	78	22

Method Validation

1. Stability: Samples from the same batch were tested at different times. The RSD of the relative retention times and relative peak areas was 2.83%, which indicated that sample solutions were stable.
2. Precision: Five consecutive injections of the samples from the same batch were conducted to determine the precision. The RSD of the relative peak areas was 0.85%, which met the requirement.
3. Reproducibility: Five samples from the same batch were prepared according to the method described above and were tested simultaneously. The RSD of the relative peak areas was 2.75%, which met the requirement.

Sample Analysis

The HPLC fingerprints of the test samples from 10 batches were analyzed under the conditions described above.

Establishment of Fingerprints

The number of peaks, peak area and relative retention time of 10 batches of water-soluble and liposoluble test samples were analyzed and compared to obtain an optimized fingerprint (Fig. 25.1).

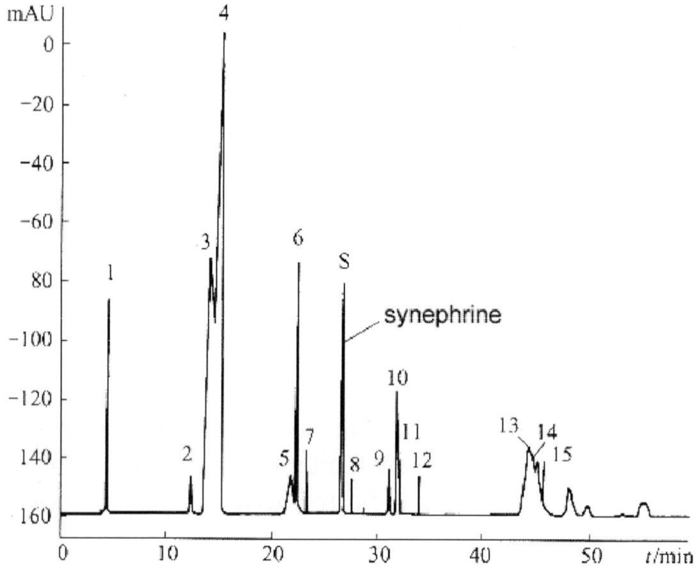

Fig. 25.1. HPLC fingerprint of Chenpi (*Pericarpium citri reticulatae*).

Results and Discussion

Common Peak Standardization

Common peaks were assigned according to the results of 10 batches of samples. There were 16 common peaks in the HPLC fingerprints of *Pericarpium citri reticulatae*. Compared to the synephrine reference peak (peak S), the relative retention times of the other 15 common peaks were 0.156, 0.451, 0.525, 0.556, 0.814, 0.826, 0.867, 1.040, 1.175, 1.198, 1.200, 1.270, 1.652, 1.685 and 1.718.

Characteristics of Common Peaks

Peaks 3 and 4 were M-shaped. Peak 5 was joined to peak 6. Peak 5 had a shoulder peak or split peak. Peaks 10 and 11 were M-shaped. Peaks 13–15 were a group of peaks. Usually peaks 13 and 14 were broad peaks and peak 15 was symmetrical.

Ratio of Common Peak Areas

Aside from the synephrine peak, no other peak area exceeded 10% of the total peak area. The total peak area of all the non-common peaks was less than 10%.

2.26

Kushen

苦参

Radix Sophorae flavescentis

Introduction

Radix Sophorae flavescentis is the dried root of *Sophora flavescens* Ait. (family Leguminosae). It originates from a lot of places in China. It is usually sold locally.

According to studies on resources, wild *Radix Sophorae flavescentis* is widely distributed in Weichang, Tangshan and Qinghuangdao of Hebei Province and the yield there is very high.

Chemical Composition

The root contains over 20 alkaloids, mainly matrine and oxymatrine. The alkaloids also include sophoranol, *N*-methylcytisine, anagyrine, baptifoline, sophocarpine, oxysophocarpine and sophoridine.

It also contains flavones, such as kurarindin, norkurarinone, kuraridinol, kurarinol, neo-kurarinol, norkurarinol, isokurarinine, formononetin, trifolirhizin and maackiain.

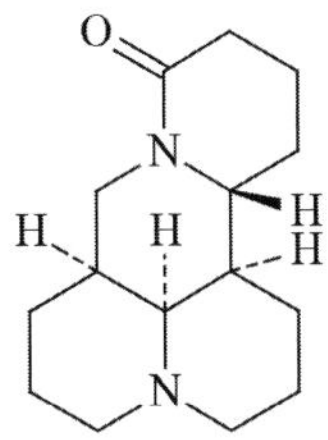

Matrine

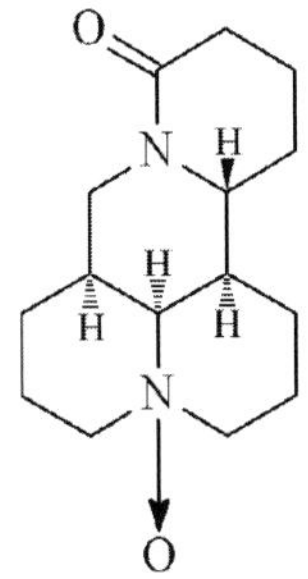

Oxymatrine

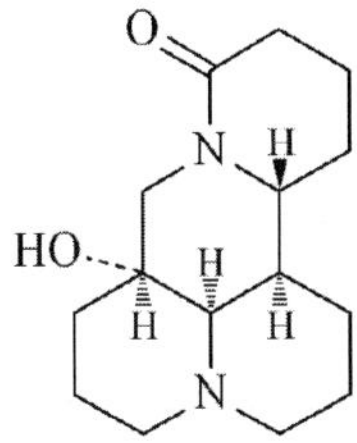

Sophoranol

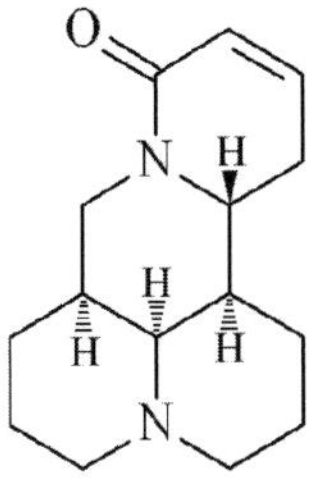

Sophocarpine

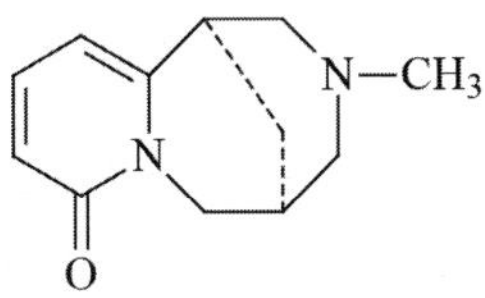

N-Methylcytisine

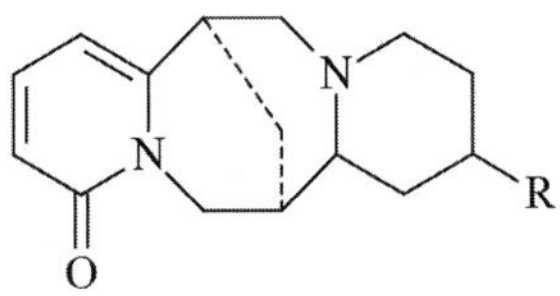

Anagyrine R=H
Baptifoline R=OH

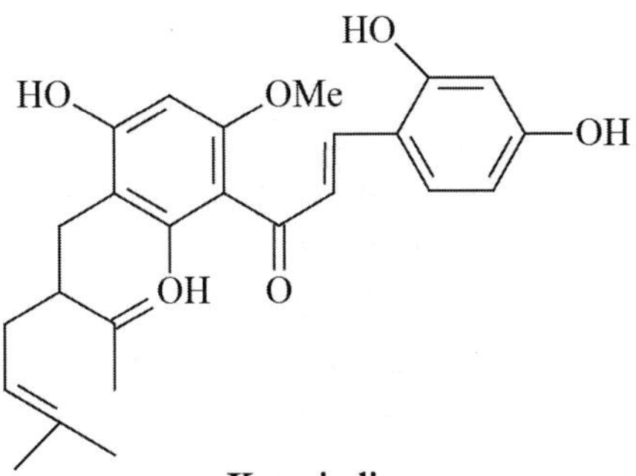

Kurarindin

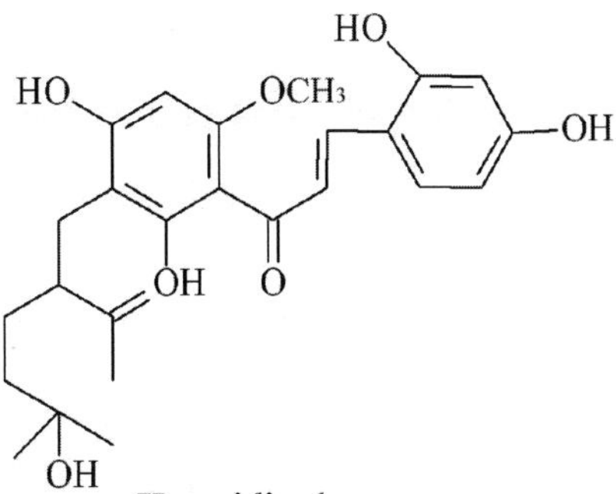

Kuraridinol

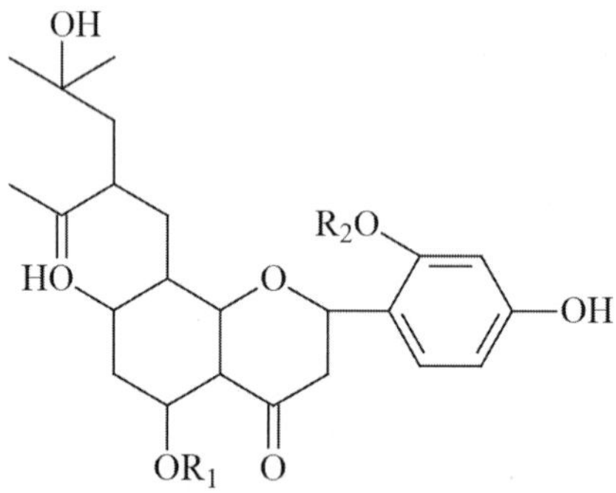

Kurarinol $R_1=CH_3$, $R_2=H$
Neo-kurarinol $R_1=R_2=CH_3$
Norkurarinol $R_1=R_2=H$

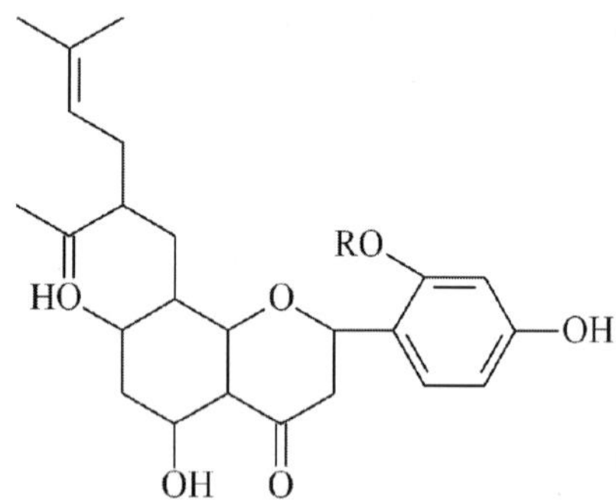

Isokurarinine $R=CH_3$
Norkurarinone $R=H$

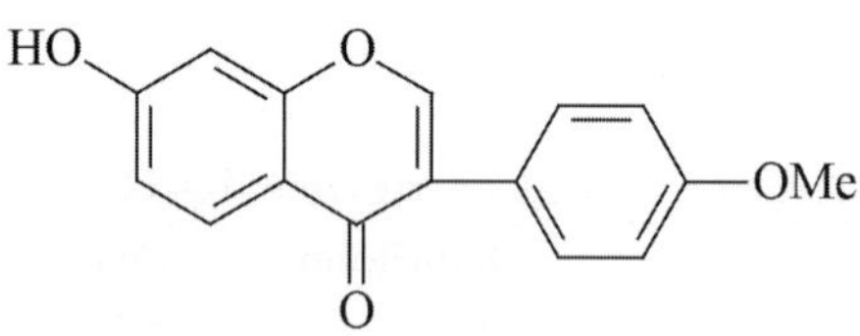

Formononetin

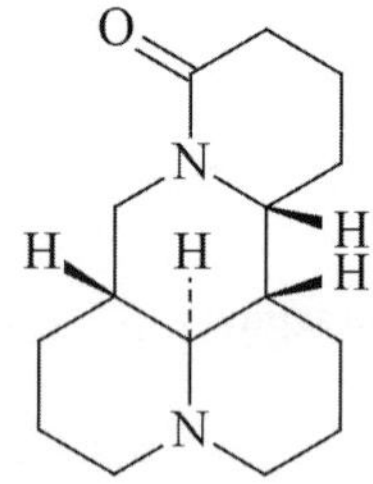

Sophoridine

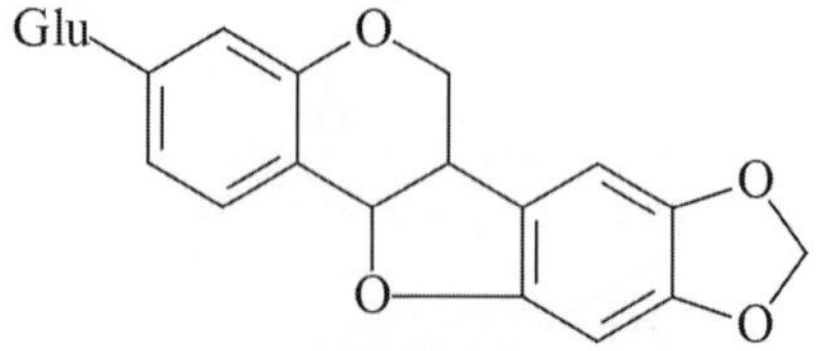

Trifolirhizin

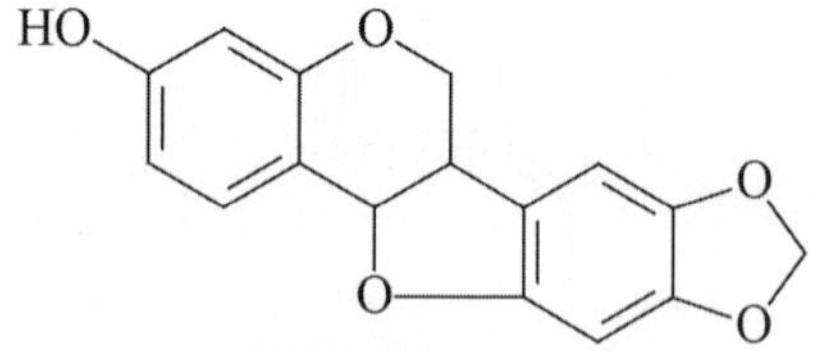

Maackiain

Experimental Materials

Samples

Wild *Radix Sophorae flavescentis* samples were collected from Weichang and Tangshan of Hebei Province in October 2002. There were 12 samples with variant identifications completed on all of them.

Reagents

Acetonitrile (chromatographic-grade, Tedia Company Inc., USA). Double-distilled deionized water (prepared by our lab). Methanol (analytical-grade methanol was purchased from Shanghai Chemical Reagents Pte. Ltd.; chromatographic-grade methanol was purchased from Tedia Company Inc.).

Preparation of Samples

2 g of crude powder of the sample was accurately weighed and placed in a round-bottom flask. 100 ml of methanol was added to the flask. After refluxing for 2 h, the sample solution was filtered. 100 ml of methanol was added to the residue and the sample was extracted for 1 h. After filtering, the two filtrates were combined. The filtrate was evaporated to dryness in an evaporating dish. The residue was dissolved and topped up to 10 ml with methanol in a volumetric flask. The mixture was filtered through a 0.45 μm Millipore filter. The filtrate was then ready to be used as a test sample.

HPLC Conditions

Instrument: Agilent 1100 high-performance liquid chromatography and diode array detector
Mobile phase: Gradient elution with water and acetonitrile
Flow rate: 1.0 ml/min
Column: Agilent ZORBAX 80A Extend-C_{18} analytical column

(250 mm × 4.6 mm, 5 μm)
Column temperature: 25°C
Detecting wavelength: 274 nm
Reference wavelength: 360 nm
Run time: 68 min
Injection volume: 10 μl

Methods

The sample solution was injected into the liquid chromatograph. The chromatogram was obtained under selected HPLC conditions. The experimental method was validated.

Mobile Phase Selection

A comparison of the experimental results showed that gradient elution with water and acetonitrile could successfully separate the chromatographic peaks and produce the maximum number of peaks from the samples. The duration of the analysis was 68 min. The 2 h chromatogram of the samples showed no characteristic peaks after 68 min.

A is water, B is acetonitrile.

	A (%)	B (%)
0–10 min	90–98	2–10
10–55 min	40–90	10–60
55–65 min	30–40	60–70
65–68 min	30–30	70–70

Method Validation

1. Stability: Test samples from the same batch were stored at room temperature and injected at regular intervals for testing. The degrees of similarity of the injections calculated by the Similarity Degree Calculation Software for Traditional Chinese Medicine Fingerprints were all above 0.92, which indicated that samples were stable within the 36 testing hours.

2. Precision: Five consecutive injections of the samples from the same batch were carried out to determine the precision. The degrees of similarity of the injections calculated by the Similarity Degree Calculation Software for Traditional Chinese Medicine Fingerprints were all above 0.94, which indicated that the injections of the test samples were carried out with precision.
3. Reproducibility: Five samples from the same batch were prepared and tested according to the method described above. The degrees of similarity of the injections calculated by the Similarity Degree Calculation Software for Traditional Chinese Medicine Fingerprints were all above 0.92, indicating good reproducibility of the results.

Sample Analysis

The HPLC fingerprints of the test samples from 10 batches were analyzed under the conditions described above.

Establishment of Fingerprints

The number of peaks, peak area and relative retention time of 10 batches of samples were analyzed and compared to obtain an optimized fingerprint (Figs. 26.1, 26.2).

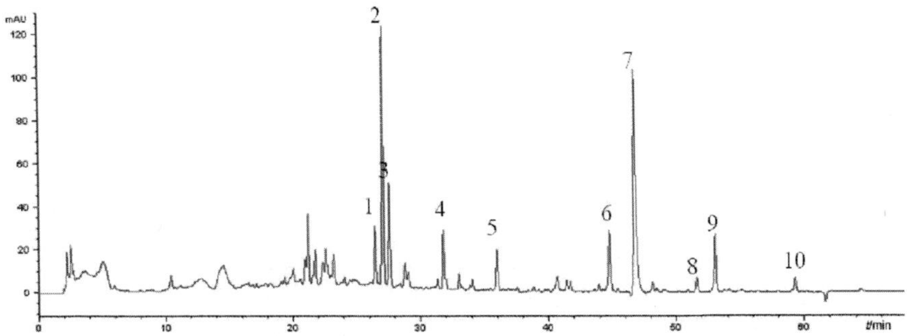

Fig. 26.1. HPLC fingerprint of Kushen (*Radix Sophorae flavescentis*).

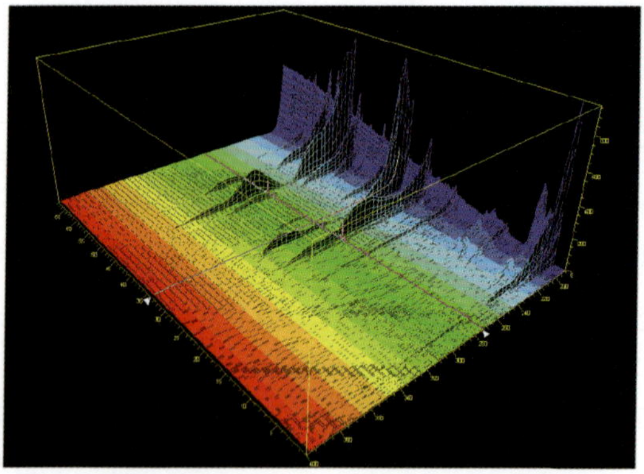

Fig. 26.2. 3-D HPLC fingerprint of Kushen (*Radix Sophorae flavescentis*).

Results and Discussion

Over 30 peaks were separated using HPLC. After comparing all the tested and documented fingerprints, 10 common peaks were selected as the characteristic peaks of the *Radix Sophorae flavescentis* fingerprint. The retention time (min) and peak area values of the 10 common peaks are listed below.

Peak number	1	2	3	4	5	6	7	8	9	10
Retention time	26.442	27.064	27.56	31.781	35.936	44.746	46.739	51.586	53.016	59.276
Peak area	273.2	1097	429.9	270.8	186	349.7	1336.6	69.3	284	80.7

Common Peak Standardization

Common peaks were assigned according to the results of 10 batches of samples. There were 10 common peaks in the HPLC fingerprints of *Radix Sophorae flavescentis*. The common peak with a retention time of about 46.74 min was subsequently used as the reference peak. Compared to the reference peak (peak 7), the relative retention times of the other nine common peaks were 0.566, 0.579, 0.590, 0.680, 0.769, 0.957, 1.104, 1.134 and 1.268.

Ratio of Common Peak Area

Aside from the reference peak, the peak area of common peak 2 exceeded 10% of the total peak area. The common peak area ratio of peak 2 was specified as 0.5745–1.0670.

2.27

Banlangen

板蓝根

Radix isatidis

Introduction

Radix isatidis is the dried root of *Isatis indigotica* Fort. (family Cruciferae). It mainly originates from Hebei and Jiangsu Provinces.

According to studies on resources, *Radix isatidis* is planted abundantly in Dingzhou, Anguo, Anhu, Yuanshi, Dingxing, Shenze, Shulu, Boye, Xinle, Gaoyi, Wangdu, Wuji, Jinzhou, Zhaoxian, Lingshou, Gaocheng, Wanquan of Hebei Province.

Chemical Composition

Radix isatidis mainly contains indigotin (0.079 mg/g), indirubin (0.0361 mg/g), 1-thiocyano-2-hydroxy-3-butene and various amino acids.

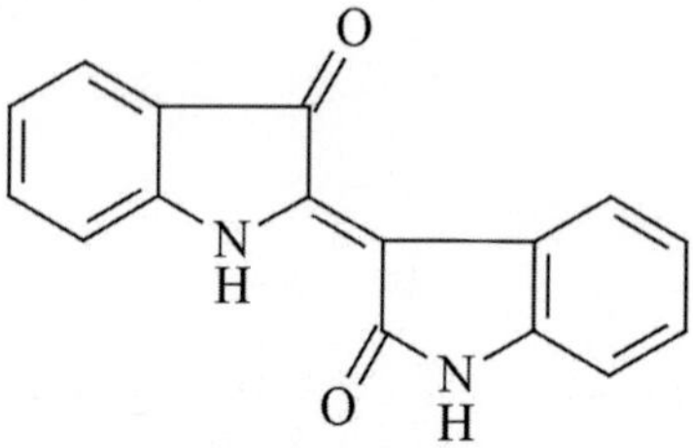

Indirubin

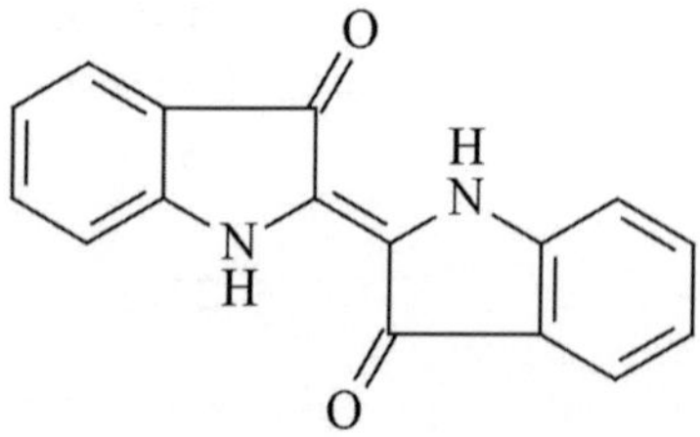

Indigotin

Experimental Materials

Samples

Radix isatidis samples were collected from one-year-old cultivated species at Anguo and Dingzhou of Hebei Province in October 2002. There were 12 samples with variant identifications completed on all of them.

Reagents

Methanol (analytical-grade methanol was purchased from Shanghai Chemical Reagents Pte. Ltd.; chromatographic-grade methanol was purchased from Tedia Company Inc.). *n*-Butanol (analytical-grade, Shanghai Chemical Reagents Pte. Ltd.). Double-distilled deionized water (prepared by our lab). Phosphoric acid (analytical-grade).

Preparation of Samples

10 g of crude powder of the sample was accurately weighed and placed in a round-bottom flask. Eight times of water was added to the flask. After being refluxed for 2 h, the extract was concentrated to 50 ml and placed in a 125 ml separating funnel. 50 ml of *n*-butanol was added. After being extracted with shaking, it was left aside to separate into different layers. The layer of *n*-butanol was collected and placed in an evaporating dish. It was evaporated to dryness over a 100°C water bath. The residue was dissolved in a small amount of methanol and filtered through a 0.45 μm Millipore filter. After being topped up to 10 ml, it was then ready to be used as a test sample.

HPLC Conditions

Instrument: Agilent 1100 high-performance liquid chromatography and diode array detector

Mobile phase: Two-step gradient elution with methanol and 0.01%

phosphoric acid
Flow rate: 1.0 ml/min
Column: HYPERSIL C_{18} analytical column (250 mm × 4.6 mm, 10 μm)
Column temperature: 30°C
Detecting wavelength: 236 nm
Reference wavelength: 360 nm
Run time: 40 min
Injection volume: 20 μl

Methods

A proper amount of sample solution was injected into the liquid chromatograph. The chromatogram was obtained under selected HPLC conditions. The experimental method was validated.

Mobile Phase Selection

A comparison of the experimental results showed that two-step gradient elution with the methanol and 0.01% phosphoric acid solution could successfully separate the chromatographic peaks and produce the maximum number of peaks from the samples. The duration of the analysis was 40 min. The 1.5 h chromatogram of the samples showed no characteristic peaks after 40 min.

A is methanol, B is 0.01% phosphoric acid.

	A (%)	B (%)
0–15 min	5–30	70–95
15–45 min	30–75	25–70
45–55 min	75–75	25–25

Method Validation

1. Stability: Test samples were stored at room temperature and tested

at 0 h, 1 h, 2 h, 4 h, 6 h, 24 h and 48 h. The degrees of similarity were analyzed by the Similarity Degree Calculation Software for Traditional Chinese Medicine Fingerprints. The results showed that the degrees of similarity were 0.9680–0.9987, which indicated that samples were stable within the 48 testing hours.

2. Precision: Five consecutive injections of the samples were conducted to determine the precision. The degrees of similarity were analyzed by the Similarity Degree Calculation Software for Traditional Chinese Medicine Fingerprints. The results showed that the degrees of similarity were 0.9695–0.9985, which indicated that the injections of the test samples were carried out with precision.
3. Reproducibility: Five samples from the same batch were prepared according to the method described above and were tested simultaneously. The degrees of similarity were analyzed by the Similarity Degree Calculation Software for Traditional Chinese Medicine Fingerprints. The results showed that the degrees of similarity were 0.9520–0.9995, indicating good reproducibility of the results.

Sample Analysis

The HPLC fingerprints of the test samples from 10 batches were analyzed under the conditions described above.

Establishment of Fingerprints

The number of peaks, peak area and relative retention time of 10 batches of samples were analyzed and compared to obtain an optimized fingerprint (Fig. 27.1).

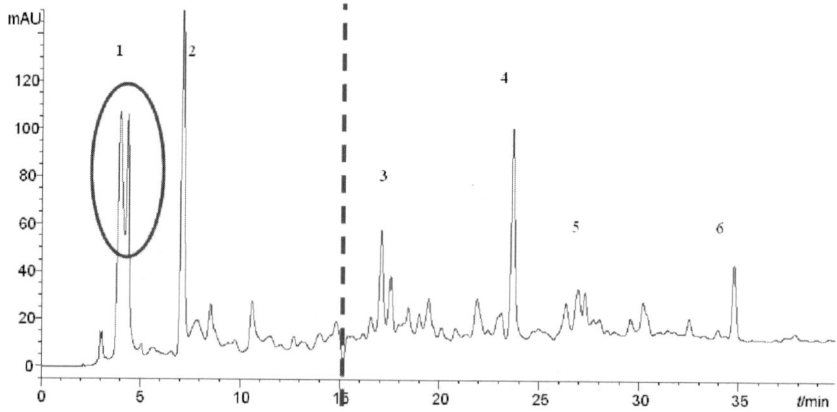

Fig. 27.1. HPLC fingerprint of Banlangen (*Radix isatidis*).

Results and Discussion

About 20 peaks were separated using HPLC. After comparing all the tested and documented fingerprints, six common peaks were selected as the characteristic peaks of *Radix isatidis*. The retention time (min) and peak area values of the six common peaks are listed below.

Peak number	1	2	3	4	5	6
Retention time	4.283	7.136	17.52	23.786	27.046	34.928
Peak area	3509.80	2124.92	981.23	1783.77	695.82	824.16

Common Peak Standardization

Common peaks were assigned according to the results of 10 batches of samples. There were six common peaks in the HPLC fingerprints of *Radix isatidis*. Peak 1 of the test samples was used as the reference peak. Compared to the reference peak, the relative retention times of the other five common peaks were 1.666, 4.091, 5.554, 6.315 and 8.155.

Characteristics of Common Peak Area

The typical characteristic was that peak 1 was a strong scissors-shaped peak.

Ratio of Common Peak Area

Aside from the reference peak, none of the other common peak areas exceeded 10% of the total peak area.

2.28

Zhimu
知母
Rhizoma anemarrhenae

Introduction

Rhizoma anemarrhenae is the dried rhizome of *Anemarrhena asphodeloides* Bge. (family Liliaceae). It is mainly produced in Hebei Province.

According to studies on resources, *Rhizoma anemarrhenae* is planted abundantly in Yixian, Anguo, Laiyuan, Chicheng, Zhangbei, Shangyi, Wanquan, Huailai, Weixian, Yangyuan, Laishui, Xuanhua, Shexian, Funing, Wu'an, Buping, Zunhua, Guyuan, Chongli, Zhuolu, Jinxing, Cixian and Xingtai of Hebei Province, of which Yixian is renown for its excellent quality of *Rhizoma anemarrhenae*.

Chemical Composition

Rhizoma anemarrhenae mainly contains sapogenins (timosaponin, mangiferin, sarsasapogenin, etc), anemaran and many metal elements.

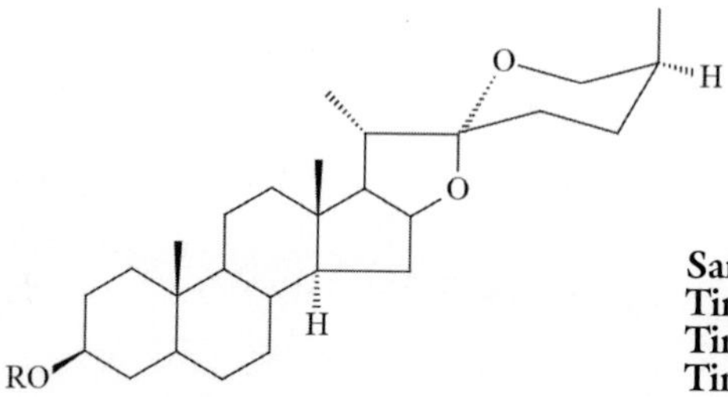

Sarsasapogenin R=H
Timosaponin A-I R=gal
Timosaponin A-III R=timobiose
Timosaponin B-I R=timobiose+glu

Experimental Materials

Samples

Rhizoma anemarrhenae samples were collected from Anguo and Yixian of Hebei Province in October 2002. There were 12 samples with variant identifications completed on all of them.

Reagents

Methanol (analytical-grade methanol was purchased from Shanghai Chemical Reagents Pte. Ltd.; chromatographic-grade methanol was purchased from Tedia Company Inc.). *n*-Butanol (analytical-grade, Shanghai Chemical Reagents Pte. Ltd.). Double-distilled deionized water (prepared by our lab). Phosphoric acid (analytical-grade).

Preparation of Samples

10 g of crude powder of the *Rhizoma anemarrhenae* sample was accurately weighed and placed in a round-bottom flask. Eight times of water was added to the flask. After being refluxed for 2 h, the extract was concentrated to 50 ml and placed in a 125 ml separating funnel. 50 ml of *n*-butanol was added to the funnel. After being extracted with shaking, it was set aside for the layers to separate. The layer of *n*-butanol was collected and placed on an evaporating dish. It was evaporated to dryness over a 100°C water bath. The residue was dissolved in a proper amount of methanol and filtered through a 0.45 μm Millipore filter. After being topped up to 10 ml, it was then ready to be used as a test sample.

HPLC Conditions

Instrument: Agilent 1100 high-performance liquid chromatography and diode array detector
Mobile phase: Two-step gradient elution with methanol and 0.01% phosphoric acid
Flow rate: 1.0 ml/min

Column: HYPERSIL C_{18} analytical column (250 mm × 4.6 mm, 10 μm)
Column temperature: 30°C
Detecting wavelength: 236 nm
Reference wavelength: 360 nm
Run time: 40 min
Injection volume: 20 μl

Methods

A proper amount of sample solution was injected into the liquid chromatograph. The chromatogram was obtained under selected HPLC conditions. The experimental method was validated.

Mobile Phase Selection

A comparison of the experimental results showed that two-step gradient elution with the methanol and 0.01% phosphoric acid solution could successfully separate the chromatographic peaks and produce the maximum number of peaks from the samples. The duration of the analysis was 40 min. The 1.5 h chromatogram of the samples showed no characteristic peaks after 40 min.

A is methanol, B is 0.01% phosphoric acid.

	A (%)	B (%)
0–15 min	5–30	70–95
15–45 min	30–75	25–70
45–55 min	75–75	25–25

Method Validation

1. Stability: Test samples were stored at room temperature and tested at 0 h, 1 h, 2 h, 4 h, 6 h, 24 h and 48 h. The degrees of similarity were analyzed by the Similarity Degree Calculation Software for Traditional Chinese Medicine Fingerprints. The results showed that the degrees of similarity were 0.9894–0.9993, which indicated that samples were stable within the 48 testing hours.

2. Precision: Five consecutive injections of the samples were conducted to determine the precision. The degrees of similarity were analyzed by the Similarity Degree Calculation Software for Traditional Chinese Medicine Fingerprints. The results showed that the degrees of similarity were 0.9806–0.9992, which indicated that the injections of the test samples were carried out with precision.
3. Reproducibility: Five samples from the same batch were prepared according to the method described above and were tested simultaneously. The degrees of similarity were analyzed by the Similarity Degree Calculation Software for Traditional Chinese Medicine Fingerprints. The results showed that the degrees of similarity were 0.9796–0.9985, indicating good reproducibility of the results.

Sample Analysis

The HPLC fingerprints of the test samples from 10 batches were analyzed under the conditions described above.

Establishment of Fingerprints

The number of peaks, peak area and relative retention time of 10 batches of samples were analyzed and compared to obtain an optimized fingerprint (Fig. 28.1).

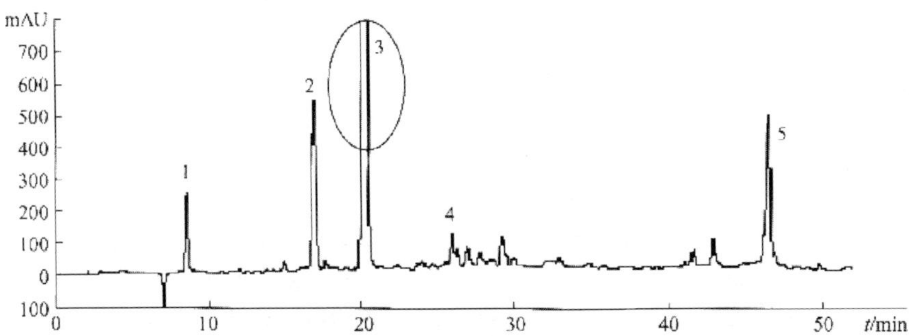

Fig. 28.1. HPLC fingerprint of Zhimu (*Rhizoma anemarrhenae*).

Results and Discussion

About 20 peaks were separated using HPLC. After comparing all the tested and documented fingerprints, five common peaks were selected as the characteristic peaks of *Rhizoma anemarrhenae*. The retention time (min) and peak area values of the five common peaks are listed below.

Peak number	1	2	3	4	5
Retention time	8.651	27.162	20.143	–	46.324
Peak area	9917.48	13051.4	33460.5	–	14116.1

Common Peak Standardization

Common peaks were assigned according to the results of 10 batches of samples. There were five common peaks in the HPLC fingerprints of *Rhizoma anemarrhenae*. Peak 3 of the test samples was used as the reference peak. Compared to the reference peak, the relative retention times of the other three common peaks (peak 4 was very weak and did not provide detailed data) were 0.429, 1.348 and 2.300.

Characteristics of Common Peak Area

The typical characteristic was that peak 3 was a particularly strong peak. Peak 2 was stronger than peak 1 and peak 3 was stronger than peak 2. Peak 4 was a group of weaker absorption peaks.

Ratio of Common Peak Area

Aside from the reference peak, none of the common peak areas exceeded 10% of the total peak area.

2.29

Cebaiye
侧柏叶
Cacumen platycladi

Introduction

Cacumen platycladi is the dried branch tip and leaf of *Platycladus orientalis* (L.) Franco (family Cupressaceae). It is a specialty from China and is produced in most parts of China.

Chemical Composition

Cacumen platycladi contains 0.75–1% of volatile oils. The main components of the volatile oils include β-thujaplicin, γ-thujaplicin, α-thujone, fenchanone, thujene, α-biotol and β-biotol.

Cacumen platycladi also contains flavonoids, which include quercetin, myricetin, kaempferide, hinokiflavone and amentoflavone.

The wax on the surface of the leaves is in the form of polyester. After saponification, 80% of the separated products are organic acids, which mainly are juniperic acid and small amounts of sabinic acid. There are also some tannins.

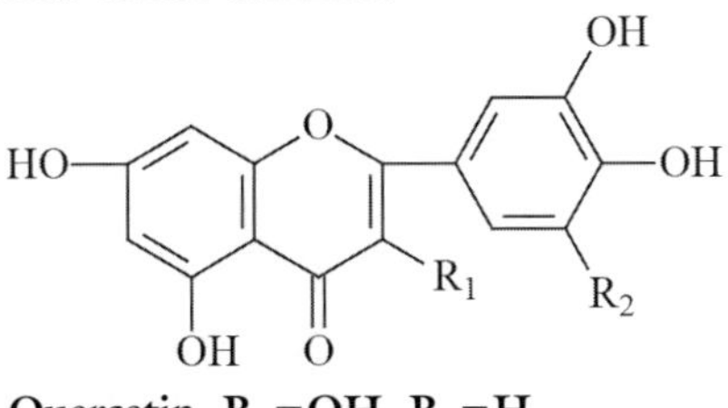

Quercetin R_1=OH, R_2=H
Quercitrin R_1=*O*-rha, R_2=H
Myricetin R_1=R_2=OH

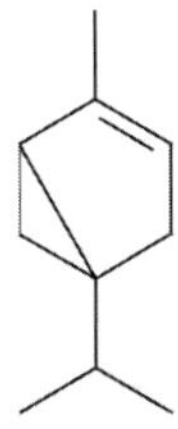

Thujene

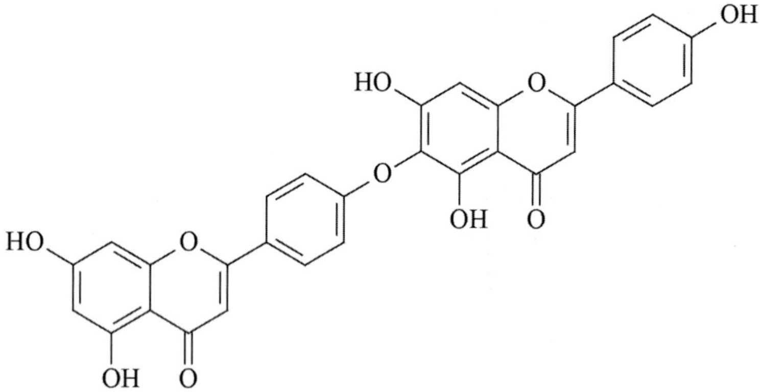

Hinokiflavone

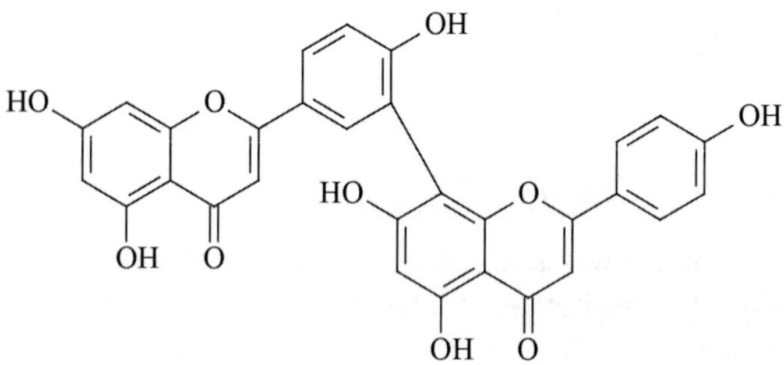

Amentoflavone

Experimental Materials

Samples

Samples 1–4 were collected from Lushanxian Zhaopingtai Reservoir of Henan Province in late September 2000. Samples 5–6 were collected from Lushanxian Sikeshu Village of Henan Province in late September 2000. Sample 7 was purchased from Haozhou Herb Market of Anhui Province in September 2000. Sample 8 was purchased from a pharmacy in Hangzhou of Zhejiang Province in late September 2000. Sample 9 was purchased from Juhuayuan Herb Market in Kunming of Yunnan Province in November 2000. Sample 10 was purchased from Anguo Herb Market of Hebei Province in

November 2000. Variant identifications were completed on all of the 10 samples.

References

Quercetin was provided by the National Institute for the Control of Pharmaceutical and Biological Products (China). Quercitrin prepared by our lab had gone through structural identification and its purity was 98.8%.

Reagents

Methanol (chromatographic-grade). Acetic acid (analytical-grade). Potassium dihydrogen phosphate (analytical-grade).

Preparation of Samples

1 g of powder sample was accurately weighed and immersed in 25 ml of methanol overnight. It was extracted by ultrasound for 30 min and filtered. The filtrate was then ready to be used as a test sample.

Preparation of References

A proper amount of quercetin and quercitrin was accurately weighed and the mixture was dissolved in methanol to form a 0.1 mg/ml solution. A proper amount of quercitrin was accurately weighed and was dissolved in methanol to form a 0.1 mg/ml solution. They were then ready to be used as references.

HPLC Conditions

Instrument: Waters 2690 HPLC Separations Module, Waters 996 Photodiode Array Detector, Millennium 32 Workstation. Rotary evaporator (BÜCHI Labortechnik AG, Switzerland)
Mobile phase: Gradient elution with methanol (A) and 0.1% glacial acetic acid in 0.01 mol/ml potassium dihydrogen phosphate (B)

Flow rate: 0.8 ml/min
Column: DikmaDiamonsil™ C_{18} (250 mm × 4.6 mm, 5 μm)
Column temperature: 25°C
Detecting wavelength: 254 nm
Run time: 60 min
Injection volume: 10 μl sample, 10 μl reference

Methods

The sample solution was injected into the liquid chromatograph. The chromatogram was obtained under selected HPLC conditions. The experimental method was validated.

Mobile Phase Selection

A comparison of the experimental results showed that when methanol and 0.1% glacial acetic acid in 0.01 mol/ml potassium dihydrogen phosphate were used as the mobile phase, gradient elution could successfully separate the chromatographic peaks and produce the maximum number of peaks. The duration of the analysis was 60 min. The 1.5 h chromatogram of the samples showed no characteristic peaks after 60 min.

A is methanol and B is 0.1% glacial acetic acid in 0.01 mol/ml potassium dihydrogen phosphate.

	A (%)	B (%)
0–55 min	15–85	15–85
55–65 min	15–85	15–85

Method Validation

1. Precision: 20 μl of accurately aspirated reference solution was injected and analyzed six times according to the HPLC conditions mentioned above. The theoretical plate number and degree of separation of the peaks were calculated. The results showed that the average theoretical plate number calculated from the quercitrin

peak was 64339 and the RSD was 2.5% (n = 6). The average theoretical plate number calculated from the quercetin peak was 68417 and the RSD was 3.8% (n = 6). The average degree of separation of the two peaks was 7.2 and the RSD was 1.8% (n = 6). These indicated that the injections of the test samples were carried out with precision.

2. Reproducibility: Six *Cacumen platycladi* samples (sample 1) were accurately weighed and prepared according to the method described above. 20 μl of test sample was injected and the quercitrin peak was used as the reference peak to analyze the relative retention times, relative peak areas and RSDs of those main peaks in six tests. The results showed that peak 7 was weak and its degrees of separation with adjacent peaks were low. The standard deviation of the relative peak areas of peak 7 was high. Aside from that, the RSDs of the other peaks were all less than 2.6%, indicating good reproducibility of the results.

Sample Analysis

The HPLC fingerprints of the test samples from 10 batches were analyzed under the conditions described above.

Establishment of Fingerprints

The number of peaks, peak area and relative retention time of 10 batches of samples were analyzed and compared to obtain an optimized fingerprint (Figs. 29.1, 29.2).

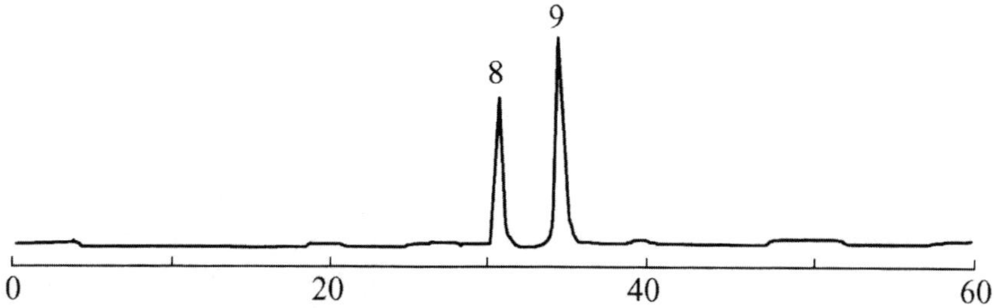

Fig. 29.1. HPLC fingerprint of references: quercitrin (8) and quercetin (9).

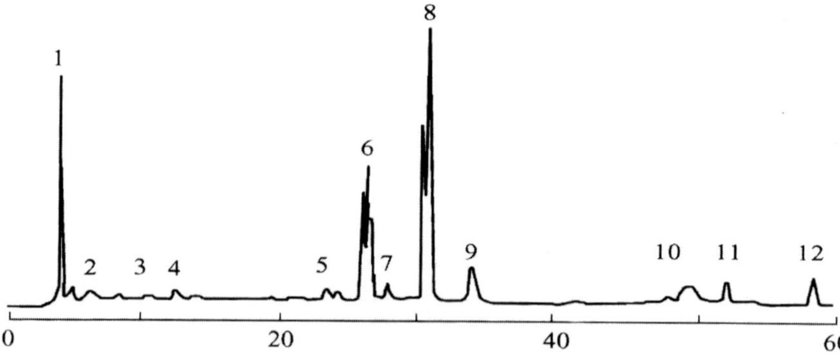

Fig. 29.2. HPLC fingerprint of Cebaiye (*Cacumen platycladi*).

Results and Discussion

Over 20 peaks were separated using HPLC. After comparing all the tested and documented fingerprints, 12 common peaks were selected as the characteristic peaks of the *Cacumen platycladi* fingerprint.

Common Peak Standardization

Common peaks were assigned according to the results of 10 batches of samples. There were 12 common peaks in the HPLC fingerprints of *Cacumen platycladi*. Compared to the quercitrin reference peak (peak 8), the relative retention times of the other 11 common peaks were 0.13, 0.21, 0.28, 0.41, 0.77, 0.86, 0.91, 1.11, 1.60, 1.69 and 1.89.

Ratio of Common Peak Areas

Aside from the quercitrin reference peak, none of the common peak areas exceeded 10% of the total peak area.

2.30

Qiancao

茜草

Radix rubiae

Introduction

Radix rubiae is the dried root and rhizome of *Rubia cordifolia* L. (family Rubiaceae). It is mainly produced in Shaanxi, Henan, Anhui, Jiangsu and Shandong Provinces.

According to studies on resources, *Radix rubiae* is planted abundantly in Liu'an, Wuhu and Ma'anshan of Anhui Province.

Chemical Composition

Radix rubiae contains anthraquinones, which are purpurin, purpuroxanthin, alizarin, munjistin and pseudopurpurin. The naphthoic acid glycosides I, II, and aglycons, which can increase the activity of leukocytes, and the rubian-ruberythric acid of alizarin are all rubia naphthoic acids.

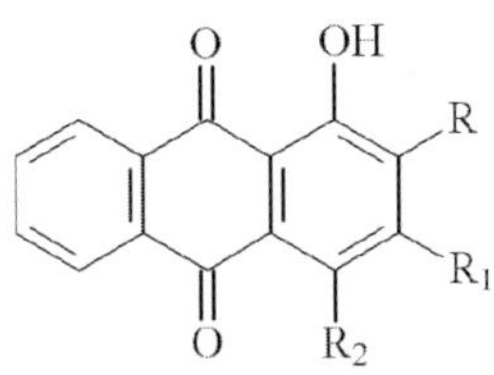

Purpurin $R=R_2=OH$, $R_1=H$

Purpuroxanthin $R=R_2=H$, $R_1=OH$

Alizarin $R_1=R_2=H$, $R=OH$

Munjistin $R=COOH$, $R_1=OH$, $R_2=H$

Pseudopurpurin $R=R_2=OH$, $R_1=COOH$

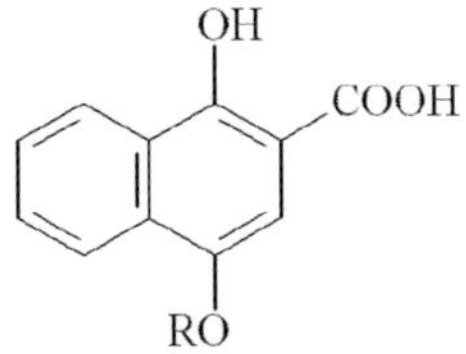

Naphthoic acid $R=H$

Naphthoic glycoside $R=glu+xyl+rha$

Experimental Materials

Samples

Radix Rubiae samples were collected in Liu'an of Anhui Province in October 2002. There were 11 samples with variant identifications completed on all of them.

Reagents

Acetonitrile (chromatographic-grade, Tedia Company Inc., USA). Double-distilled deionized water (prepared by our lab). Methanol (analytical-grade methanol was purchased from Shanghai Chemical Reagents Pte. Ltd.; chromatographic-grade methanol was purchased from Tedia Company Inc.).

Preparation of Samples

2 g of crude powder of the sample was accurately weighed and placed in a round-bottom flask. 100 ml of methanol was added to the flask. After 2 h of extraction by reflux, the sample solution was filtered. 100 ml of methanol was added to the residue and the solution was extracted for another 1 h. After filtering, the two filtrates were combined. The filtrate was evaporated to dryness on an evaporating dish. The residue was topped up to 10 ml with methanol in a volumetric flask. The mixture was filtered through a 0.45 μm Millipore filter. The filtrate was then ready to be used as a test sample.

HPLC Conditions

Instrument: Agilent 1100 high-performance liquid chromatography and diode array detector
Mobile phase: Water and acetonitrile with gradient elution
Flow rate: 1.0 ml/min
Column: Agilent ZORBAX 80A Extend-C_{18} analytical column (250 mm × 4.6 mm, 5 μm)

Column temperature: 25°C
Detecting wavelength: 270 nm
Reference wavelength: 360 nm
Run time: 45 min
Injection volume: 10 μl

Methods

The sample solution was injected into the liquid chromatograph. The chromatogram was obtained under selected HPLC conditions. The experimental method was validated.

Mobile Phase Selection

A comparison of the experimental results showed that gradient elution with water and acetonitrile could successfully separate the chromatographic peaks and produce the maximum number of peaks from the samples. The duration of the analysis was 45 min. The 1.5 h chromatogram of the samples showed no characteristic peaks after 45 min.

A is water, B is acetonitrile.

	A (%)	B (%)
0–10 min	72–98	2–28
10–40 min	15–72	28–85
40–45 min	15–15	85–85

Method Validation

1. Stability: Test samples from the same batch were stored at room temperature and injected at regular intervals for testing. The degrees of similarity of the injections calculated by the Similarity Degree Calculation Software for Traditional Chinese Medicine Fingerprints were all above 0.95, which indicated that samples were stable within the 36 testing hours.

2. Precision: Five consecutive injections of the samples from the same batch were carried out to determine the precision. The degrees of similarity of the injections calculated by the Similarity Degree Calculation Software for Traditional Chinese Medicine Fingerprints were all above 0.96, which indicated that the injections of the test samples were carried out with precision.
3. Reproducibility: Five samples from the same batch were prepared and tested according to the method described above. The degrees of similarity of the injections calculated by the Similarity Degree Calculation Software for Traditional Chinese Medicine Fingerprints were all above 0.94, indicating good reproducibility of the results.

Sample Analysis

The HPLC fingerprints of the test samples from 10 batches were analyzed under the conditions described above.

Establishment of Fingerprints

The number of peaks, peak area and relative retention time of 10 batches of samples were analyzed and compared to obtain an optimized fingerprint (Figs. 30.1, 30.2).

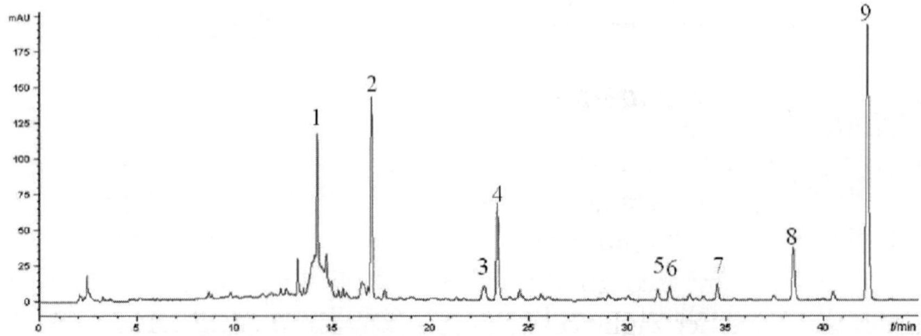

Fig. 30.1. HPLC fingerprint of Qiancao (*Radix rubiae*).

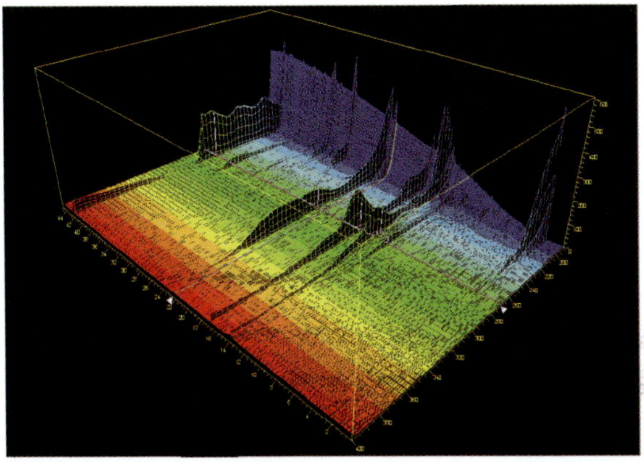

Fig. 30.2. 3-D HPLC fingerprint of Qiancao (*Radix rubiae*).

Results and Discussion

Over 30 peaks were separated using HPLC. After comparing all the tested and documented fingerprints, nine common peaks were selected as the characteristic peaks of the *Radix rubiae* fingerprint. The retention time (min) and peak area values of the nine common peaks are listed below.

Peak number	1	2	3	4	5	6	7	8	9
Retention time	13.214	16.991	22.689	23.359	31.504	32.102	34.541	38.45	42.211
Peak area	203.6	962.4	146.4	563.6	70.1	102.3	116.8	367.8	1820.3

Common Peak Standardization

Common peaks were assigned according to the results of 10 batches of samples. There were nine common peaks in the HPLC fingerprints of *Radix rubiae*. Common peak 9 was subsequently used as the reference peak. Compared to the reference peak, the relative retention times of the other eight common peaks were 0.313, 0.403, 0.538, 0.553, 0.746, 0.761, 0.818 and 0.911.

Ratio of Common Peak Area

Aside from the reference peak, the peak area of common peak 2 exceeded 10% of the total peak area and its relative peak area ratio was specified as 0.3701–0.6873.

2.31

Nanshashen
南沙参
Radix adenophorae

Introduction

Radix adenophorae is the dried root of *Adenophora tetraphylla* (Thunb.) Fisch. or *Adenophora stricta* Miq. (family Campanulaceae). It is mainly produced in Anhui, Jiangsu, Zhejiang and Guizhou Provinces.

According to studies on resources, *Radix adenophorae* is planted abundantly in Nanjing, Pukou (Jiangpu), Jiangning, Jurong, Dantu and Danyang of Jiangsu Province, and Ma'anshan and Hexian of Anhui Province.

Chemical Composition

Radix adenophorae contains nanshashen saponin, β-sitosterol, daucosterol, taraxerone, saturated fatty acids, phospholipids, γ-terpinene, camphene, eucalyptol, camphor and bornyl acetate. It also contains xanthotoxin (8-methoxypsoralen), which is a component of furocoumarins.

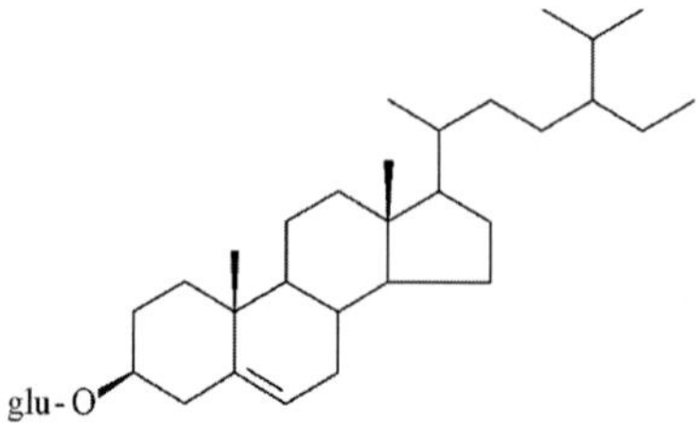

Daucosterol

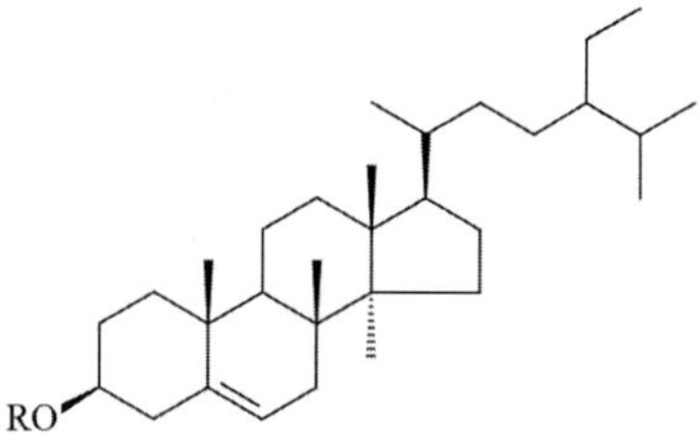

β**-Sitosterol R=H**

β**-Sitosterol-**α**-glucoside R=glu**

Experimental Materials

Samples

Radix adenophorae samples were collected from Nanjing of Jiangsu Province in late September 2002. There were 12 samples with variant identifications completed on all of them.

Reagents

Acetonitrile (chromatographic-grade, Tedia Company Inc., USA). Double-distilled deionized water (prepared by our lab). Methanol (analytical-grade methanol was purchased from Shanghai Chemical Reagents Pte. Ltd.; chromatographic-grade methanol was purchased from Tedia Company Inc.).

Preparation of Samples

2 g of crude powder of the sample was accurately weighed and placed in a round-bottom flask. 100 ml of methanol was added to the flask. After 2 h of extraction by reflux, the sample solution was filtered. 100 ml of methanol was added to the residue and the solution was extracted for another 1 h. After filtering, the two filtrates were combined. The

filtrate was evaporated to dryness on an evaporating dish. The residue was dissolved and topped up to 10 ml with methanol in a volumetric flask. The mixture was filtered through a 0.45 μm Millipore filter. The filtrate was then ready to be used as a test sample.

HPLC Conditions

Instrument: Agilent 1100 high-performance liquid chromatography and diode array detector
Mobile phase: Water and acetonitrile with gradient elution
Flow rate: 1.0 ml/min
Column: Agilent ZORBAX 80A Extend-C_{18} analytical column (250 mm × 4.6 mm, 5 μm)
Column temperature: 25°C
Detecting wavelength: 202 nm
Reference wavelength: 360 nm
Run time: 70 min
Injection volume: 10 μl

Methods

The sample solution was injected into the liquid chromatograph. The chromatogram was obtained under selected HPLC conditions. The experimental method was validated.

Mobile Phase Selection

A comparison of the experimental results showed that gradient elution with water and acetonitrile could successfully separate the chromatographic peaks and produce the maximum number of peaks from the samples. The duration of the analysis was 70 min. The 2 h chromatogram of the samples showed no characteristic peak after 70 min.

A is water, B is acetonitrile.

	A (%)	B (%)
0–15 min	90–98	2–10
15–30 min	55–90	10–45
30–40 min	35–55	45–65
40–60 min	20–35	65–80
65–70 min	0–20	80–100

Method Validation

1. Stability: Test samples from the same batch were stored at room temperature and injected at regular intervals for testing. The degrees of similarity of the injections calculated by the Similarity Degree Calculation Software for Traditional Chinese Medicine Fingerprints were all above 0.9, which indicated that samples were stable within the 36 testing hours.
2. Precision: Five consecutive injections of the samples from the same batch were carried out to determine the precision. The degrees of similarity of the injections calculated by the Similarity Degree Calculation Software for Traditional Chinese Medicine Fingerprints were all above 0.96, which indicated that the injections of the test samples were carried out with precision.
3. Reproducibility: Five samples from the same batch were prepared and tested according to the method described above. The degrees of similarity of the injections calculated by the Similarity Degree Calculation Software for Traditional Chinese Medicine Fingerprints were all above 0.92, indicating good reproducibility of the results.

Sample Analysis

The HPLC fingerprints of the test samples from 10 batches were analyzed under the conditions described above.

Establishment of Fingerprints

The number of peaks, peak area and relative retention time of 10 batches of samples were analyzed and compared to obtain an optimized fingerprint (Figs. 31.1, 31.2).

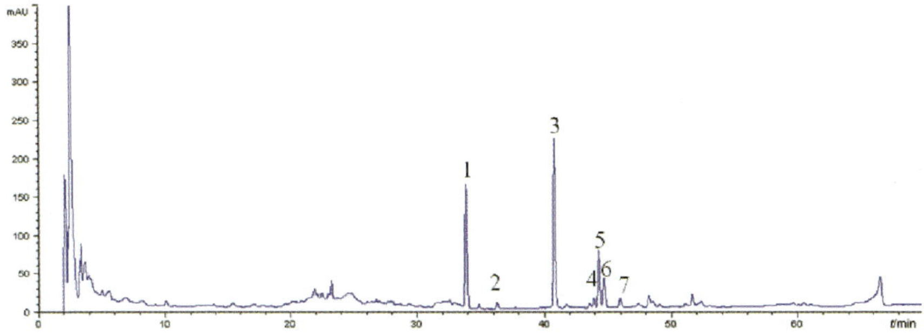

Fig. 31.1. HPLC fingerprint of Nanshashen (*Radix adenophorae*).

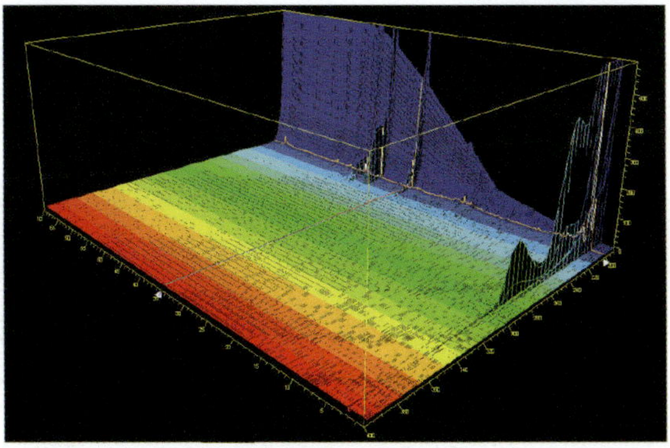

Fig. 31.2. 3-D HPLC fingerprint of Nanshashen (*Radix adenophorae*).

Results and Discussion

About 10 peaks were separated using HPLC. After comparing all the tested and documented fingerprints, seven common peaks were selected as the characteristic peaks of the *Radix adenophorae* fingerprint. The retention time (min) and peak area values of the seven common peaks are listed below.

Peak number	1	2	3	4	5	6	7
Retention time	33.805	36.259	40.754	43.849	44.232	44.651	45.917
Peak area	1871.9	100.1	2174.1	138.9	783.3	506.3	162.7

Common Peak Standardization

Common peaks were assigned according to the results of 10 batches of samples. There were seven common peaks in the HPLC fingerprints of *Radix adenophorae*. The common peak of peak 3 was subsequently used as the reference peak. Compared to the reference peak, the relative retention times of the other six common peaks were 0.829, 0.890, 1.076, 1.085, 1.096 and 1.127.

Ratio of Common Peak Area

No common peak area exceeded 10% of the total peak area.

2.32

Zhizi

栀子

Fructus gardeniae

Introduction

Fructus gardeniae is the dried mature fruit of *Gardenia jasminoides* Ellis (family Rubiaceae). It is mainly produced in Hunan, Jiangxi, Fujian, Zhejiang, Sichuan and Hubei Provinces. Hunan has the largest abundance of *Fructus gardeniae* and Zhejiang's *Fructus gardeniae* is of better quality.

According to studies on resources, *Fructus gardeniae* was produced in large scale in Jiangxi Province and the yield is relatively stable. There are seven planting bases in Jiangxi, which are Tiannan, Qingxi, Damiao and Luxi of Huangchengzhen in Fengcheng City, Shuime Village of Jinchuanzhen in Xinganxian, Yichengzhen Cultivation Farm of Zhangshu City, and Chongluxiang of Jinxixian. It is also produced in Gaole, Le'an, Dongxiang and Nancheng.

Chemical Composition

Fructus gardeniae contains various iridoid glycosides, such as geniposide, gardenoside, genipin-1-β-D-gentiobioside, shanzhiside, gardoside, scandoside methyl ester, deacetyl-asperulosidic acid methyl ester and geniposidic acid. It also contains gardenin, crocin and crocetin.

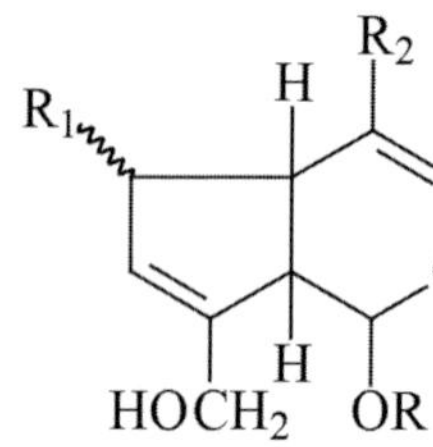

Geniposide R=glu, R_1=H, R_2=$COOCH_3$
Genipin-1-β-D-gentiobioside R=gentiobiose, R_1=H, R_2=$COOCH_3$
Scandoside methyl ester R=glu, R_1=β-OH, R_2=$COOCH_3$
Deacetyl-asperulosidic acid methyl ester R=glu, R_1=α-OH, R_2=$COOCH_3$
Geniposidic acid R=glu, R_1=H, R_2=COOH

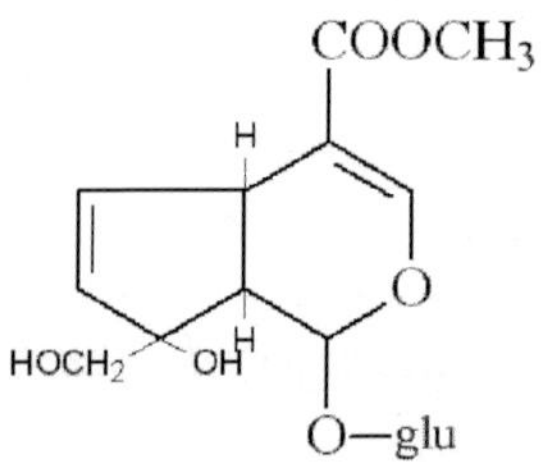

Gardenoside

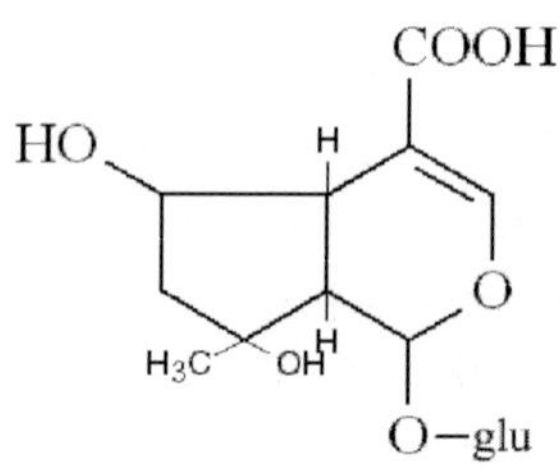

Shanzhiside

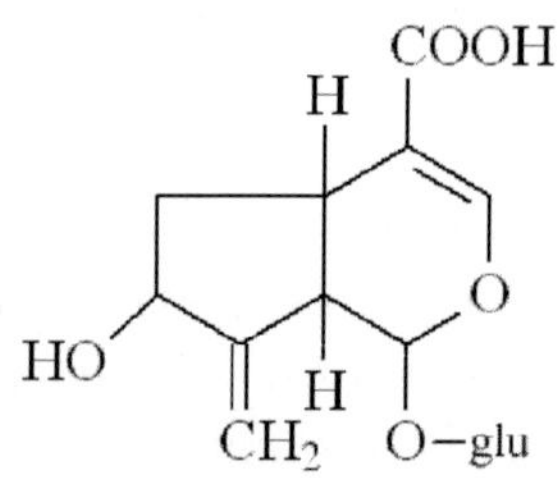

Gardoside

Experimental Materials

Samples

Fructus gardeniae samples were cultivated species and were collected from Tiannan, Qingxi, Damiao and Luxi of Huangchengzhen in Fengcheng of Jiangxi Province in November 2001. There were 12 samples with variant identifications completed on all of them.

Reference

Geniposide used for quantitative analysis was provided by the National Institute for the Control of Pharmaceutical and Biological Products (China).

Reagents

Acetonitrile (chromatographic-grade, Tedia Company Inc., USA). Methanol (analytical-grade methanol was purchased from Shanghai Chemical Reagents Pte. Ltd.; chromatographic-grade methanol was purchased from Tedia Company Inc.). Glacial acetic acid (analytical-grade, Nanjing Chemical Reagent Co. Ltd.). Double-distilled deionized water (prepared by our lab).

Preparation of Samples

0.5 g of powder sample (passed through a 20 mesh sieve) was accurately weighed. After adding 20 ml of methanol, it was refluxed for 1 h and then filtered. The residue was refluxed two more times by the same method and was filtered again. The filtrates were combined. The solvent was evaporated to dryness at low heat. The residue was dissolved in methanol and was topped up to 25 ml. The solution was mixed well and filtered through a 0.45 μm Millipore filter. After discarding the first 5 ml of filtrate, the remaining filtrate was ready to be used as a test sample.

Preparation of Reference

A proper amount of geniposide control was accurately weighed and dissolved in methanol to make a 20 μg/ml solution. It was then ready to be used as a reference.

HPLC Conditions

Instrument: Agilent 1100 high-performance liquid chromatography and diode array detector
Column: Inertsil® ODS-3 analytical column (250 mm x 4.6 mm, 5 μm)
Mobile phase: Gradient elution with 1% HAc-H_2O and 1% HAc-CH_3CN

Flow rate: 1.0 ml/min
Column temperature: 30°C
Detecting wavelength: 254 nm
Reference wavelength: 360 nm
Run time: 80 min
Injection volume: 10 μl sample, 10 μl reference

Methods

A proper amount of sample solution was injected into the liquid chromatograph. The chromatogram was obtained under selected HPLC conditions. The experimental method was validated.

Mobile Phase Selection

A comparison of the experimental results showed that gradient elution with 1% HAc-H_2O and 1% HAc-CH_3CN could successfully separate the chromatographic peaks and produce the maximum number of peaks from the samples. The duration of the analysis was 80 min. The 2 h chromatogram of the samples showed no characteristic peaks after 80 min.

A is 1% HAc-H_2O, B is 1% HAc-CH_3CN.

	A (%)	B (%)
0–20 min	94–94	6–6
20–50 min	70–94	6–30
50–60 min	70–70	30–30

Method Validation

1. Stability: Test samples from the same batch were stored at room temperature and were tested at 0 h, 2 h, 4 h, 7 h, 11 h, 16 h, 22 h, 29 h and 36 h. The degrees of similarity calculated by the Similarity Degree Calculation Software for Traditional Chinese Medicine Fingerprints (DEMO version) were all above 0.95, which indicated that samples were stable within the 36 testing hours.

2. Precision: Five consecutive injections of the samples (from the same batch) were conducted to determine the precision. The degrees of similarity calculated by the Similarity Degree Calculation Software for Traditional Chinese Medicine Fingerprints (DEMO version) were all above 0.99, which indicated that the injections of the test samples were carried out with precision.
3. Reproducibility: Five samples from the same batch were prepared according to the method described above and were tested simultaneously. The degrees of similarity calculated by the Similarity Degree Calculation Software for Traditional Chinese Medicine Fingerprints were all above 0.95, indicating good reproducibility of the results.

Sample Analysis

The HPLC fingerprints of the test samples from 10 batches were analyzed under the conditions described above.

Establishment of Fingerprints

The number of peaks, peak area and relative retention time of 10 batches of samples were analyzed and compared to obtain an optimized fingerprint (Fig. 32.1).

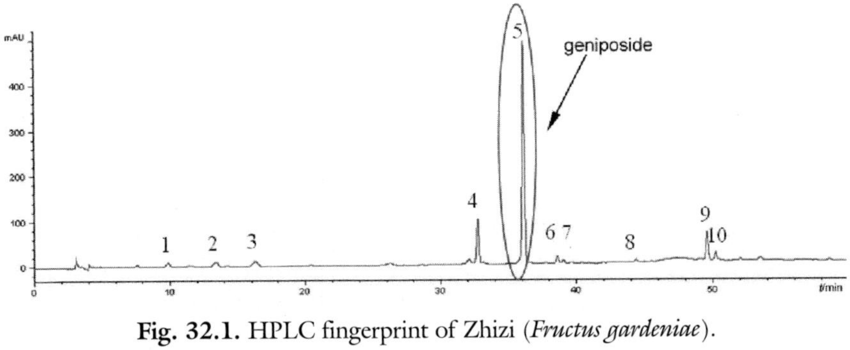

Fig. 32.1. HPLC fingerprint of Zhizi (*Fructus gardeniae*).

Results and Discussion

After comparing all the tested and documented fingerprints, 10 common peaks were selected as the characteristic peaks of *Fructus gardeniae*. The retention time (min) and peak area values of the 10 common peaks are listed below.

Peak number	1	2	3	4	5	6	7	8	9	10
Retention time	9.81	13.373	16.255	32.707	36.08	38.575	39.025	44.344	49.525	50.165
Peak area	157.9	260.2	315.4	1251	6135.3	303.4	157.5	92.6	788.2	321.7

Common Peak Standardization

Common peaks were assigned according to the results of 10 batches of samples. There were 10 common peaks in the HPLC fingerprints of *Fructus gardeniae*. Compared to the geniposide reference peak (peak 5), the relative retention times of the other nine common peaks were 0.272, 0.371, 0.451, 0.907, 1.069, 1.082, 1.229, 1.373 and 1.390.

Ratio of Common Peak Area

Aside from the geniposide reference peak (peak 5), the peak area of common peak 4 exceeded 10% of the total peak area and its relative peak area ratio was specified as 0.143–0.265.

2.33

Jiegeng
桔梗
Radix platycodonis

Introduction

Radix platycodonis is the dried root of *Platycodon grandiflorum* (Jacq.) A. DC. (family Campanulaceae). It is produced in most parts of China. The yields in Northeast or Northern China are higher. The quality of *Radix platycodonis* from Northern China is better than that from other places.

According to studies on resources, *Radix platycodonis* is planted abundantly in Nanjing, Jiangpu, Jiangning, Jurong, Sushui, Zhenjiang, Jintan, Gaochun, Yixing, Suyang, Lianyungang and Xuyi of Jiangsu Province.

Chemical Composition

The root contains various saponins, such as platycodins A, C and D. Complete hydrolysis of these saponins will produce polygalacic acid, platycodigenin and platycogenic acids A, B and C. Partial hydrolysis of these saponins will produce prosaponins, such as 3-*O*-β-D-glucopyranosyl platycodigenin, laminaribiosyl platycodigenin and gluctinobiosyl platycodigenin. It also contains platycodins A, C, D_1 and D_2.

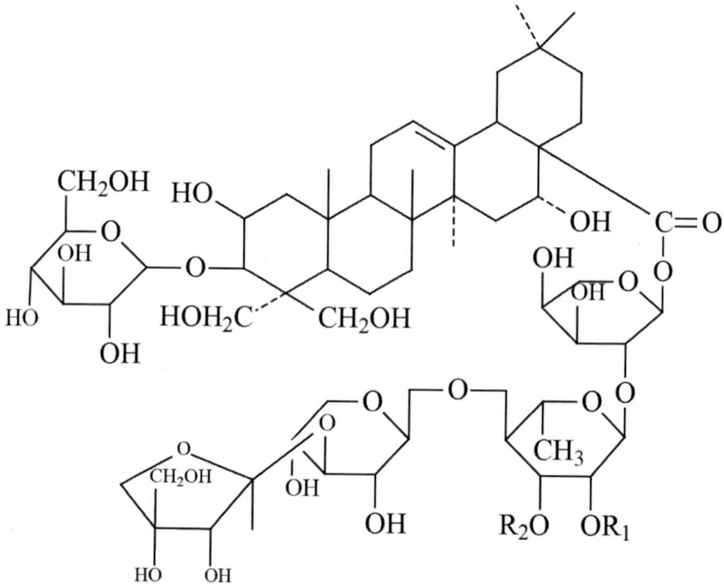

Platycodin A $R_1=COCH_3$, $R_2=H$
Platycodin C $R_1=H$, $R_2=COCH_3$
Platycodin D $R_1=R_2=H$

Experimental Materials

Samples

Radix platycodonis samples were collected from Nanjing and Yixing of Jiangsu Province in October 2002. There were 12 samples with variant identifications completed on all of them.

Reagents

Acetonitrile (chromatographic-grade, Tedia Company Inc., USA). Double-distilled deionized water (prepared by our lab). Methanol (analytical-grade methanol was purchased from Shanghai Chemical Reagents Pte. Ltd.; chromatographic-grade methanol was purchased from Tedia Company Inc.). Ether (analytical-grade, Shanghai Chemical Reagents Pte. Ltd.).

Preparation of Samples

4 g of crude powder of the sample was accurately weighed and placed in a round-bottom flask, then placed in a Soxhlet extractor. It was then immersed in 25 ml of methanol for 15 h. After adding another 25 ml of methanol, it was refluxed for 6 h. The sample solution was allowed to stand overnight and then filtered. The filtrate was concentrated to 15–20 ml over a water bath. After it had been cooled down, 50 ml of ether was added and the mixture was agitated. It was allowed to stand until it became clear. The supernatant on top was discarded and the precipitate was dissolved by adding methanol stepwise (20 ml, 10 ml and 5 ml), followed by heating each time. The mixture was cooled down and filtered. The filtrate was combined and evaporated to dryness over a water bath. It was then dissolved and topped up to 25 ml by adding methanol in a volumetric flask. After being filtered through a 0.45 μm Millipore filter, it was then ready to be used as a test sample.

HPLC Conditions

Instrument: Agilent 1100 high-performance liquid chromatography and diode array detector
Mobile phase: Gradient elution with water and acetonitrile
Flow rate: 1.0 ml/min
Column: Agilent ZORBAX 80A Extend-C_{18} analytical column (250 mm × 4.6 mm, 5 μm)
Column temperature: 25°C
Detecting wavelength: 202 nm
Reference wavelength: 360 nm
Run time: 70 min
Injection volume: 20 μl

Methods

The sample solution was injected into the liquid chromatograph. The chromatogram was obtained under selected HPLC conditions. The experimental method was validated.

Mobile Phase Selection

A comparison of the experimental results showed that gradient elution with water and acetonitrile could successfully separate the chromatographic peaks and produce the maximum number of peaks from the samples. The duration of the analysis was 70 min. The 2 h chromatogram of the samples showed no characteristic peaks after 70 min.

A is water, B is acetonitrile.

	A (%)	B (%)
0–20 min	85–98	2–15
20–40 min	65–85	15–35
40–60 min	5–65	35–95
60–70 min	5–5	95–95

Method Validation

1. Stability: Test samples from the same batch were stored at room temperature and injected at regular intervals for testing. The degrees of similarity of the injections calculated by the Similarity Degree Calculation Software for Traditional Chinese Medicine Fingerprints were all above 0.93, which indicated that samples were stable within the 36 testing hours.
2. Precision: Five consecutive injections of the samples from the same batch were carried out to determine the precision. The degrees of similarity of the injections calculated by the Similarity Degree Calculation Software for Traditional Chinese Medicine Fingerprints were all above 0.99, which indicated that the injections of the test samples were carried out with precision.
3. Reproducibility: Five samples from the same batch were prepared and tested according to the method described above. The degrees of similarity of the injections calculated by the Similarity Degree Calculation Software for Traditional Chinese Medicine Fingerprints were all above 0.95, indicating good reproducibility of the results.

Sample Analysis

The HPLC fingerprints of the test samples from 10 batches were analyzed under the conditions described above.

Establishment of Fingerprints

The number of peaks, peak area and relative retention time of 10 batches of samples were analyzed and compared to obtain an optimized fingerprint (Figs. 33.1, 33.2).

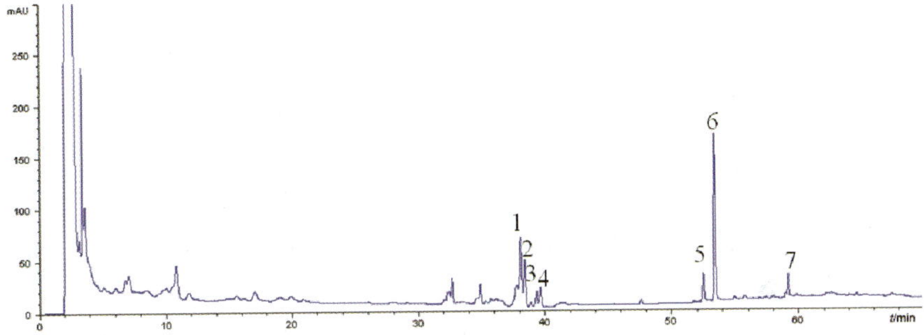

Fig. 33.1. HPLC fingerprint of Jiegeng (*Radix platycodonis*).

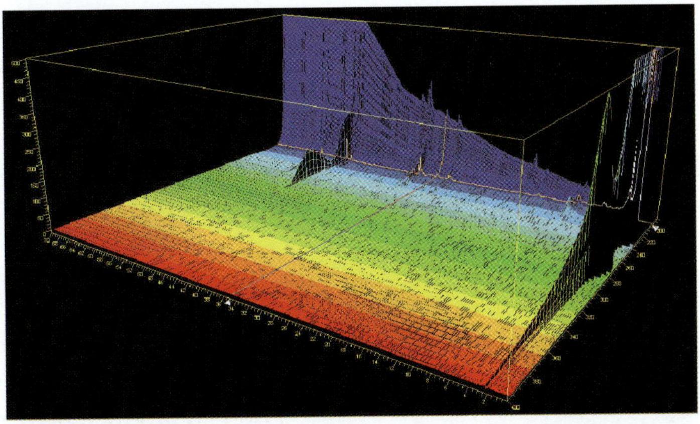

Fig. 33.2. 3-D HPLC fingerprint of Jiegeng (*Radix platycodonis*).

Results and Discussion

About 20 peaks were separated using HPLC. After comparing all the tested and documented fingerprints, seven common peaks were selected as the characteristic peaks of the *Radix platycodonis* fingerprint. The retention time (min) and peak area values of the seven common peaks are listed below.

Peak number	1	2	3	4	5	6	7
Retention time	38.054	38.395	39.306	39.612	52.467	53.362	59.139
Peak area	1252.1	452.5	161.5	206.6	301.6	1339.7	334.1

Common Peak Standardization

Common peaks were assigned according to the results of 10 batches of samples. There were seven common peaks in the HPLC fingerprints of *Radix platycodonis*. The common peak of peak 6 was subsequently used as the reference peak. Compared to the reference peak, the relative retention times of the other six common peaks were 0.713, 0.720, 0.737, 0.742, 0.983 and 1.108.

Ratio of Common Peak Area

No common peak area exceeded 10% of the total peak area.

2.34

Shegan

射干

Rhizoma belamcandae

Introduction

Rhizoma belamcandae is the dried rhizome of *Belamcanda chinensis* (L.) DC. (family Iridaceae). It is mainly produced in Henan, Hubei, Jiangsu and Anhui Provinces.

According to studies on resources, *Rhizoma belamcandae* is planted abundantly in Nanjing, Jiangning, Jiangpu, Gaochun and Jurong of Jiangsu Province.

Chemical Composition

Rhizoma belamcandae contains isoflavones, triterpenes, phytosterins and diphenylethenes. Their main active ingredients are listed below.

1. Isoflavonoids: tectoridin, iridin, irigenin, irisflorentin, tectorigenin, methyirisolidone, iridtectorigenin, 3′,5,7-trihydroxy,-4,6′ -dimethoxy isoflavone, dichotoitin, noririsflorentin, 3′,4′,5,7 tetrahydroxy-8-methoxy isoflavone, shegansu A, irigenin-5-*O*-(6′-*O*-vanillic acid)-*β*-D-glucopyranoside.
2. Stilbenes: resveratrol, isorhapontigenin, bisisorhapontigenin.

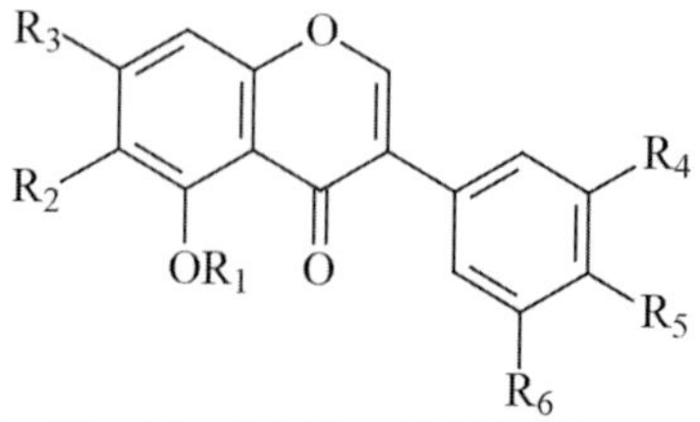

Tectoridin $R_1=R_4=R_6=H$, $R_2=OCH_3$, $R_3=O$-glu, $R_5=OH$
Iridin $R_1=H$, $R_2=R_4=R_5=R_6=OCH_3$, $R_3=O$-glu
Irigenin $R_1=H$, $R_2=R_5=R_6=OCH_3$, $R_3=R_4=OH$
Irisflorentin $R_1=CH_3$, $R_2=R_3=O$-CH_2-O, $R_4=OH$, $R_5=R_6=OCH_3$
Tectorigenin $R_1=R_3=R_6=H$, $R_2=OCH_3$, $R_4=R_5=OH$
Dichotoitin $R_1=H$, $R_2=R_3=O$-CH_2-O, $R_4=OH$, $R_5=R_6=OCH_3$

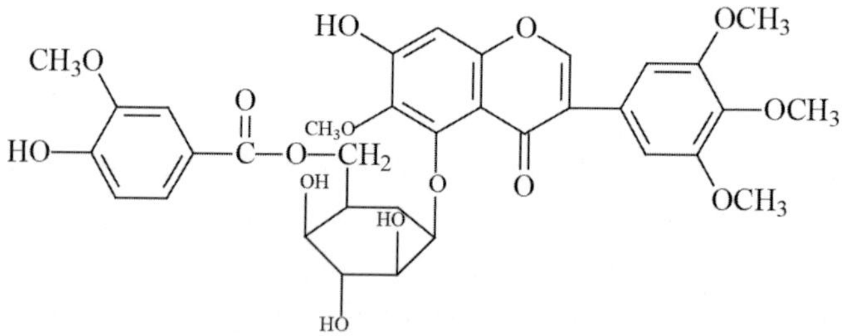

Shegansu A

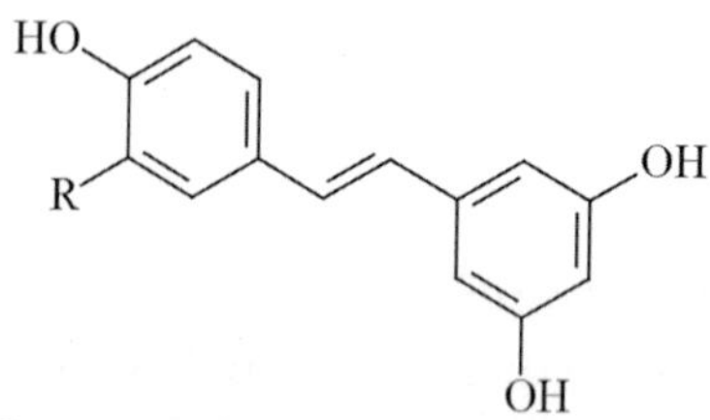

Resveratrol R=OH
Isorhapontigenin $R=OCH_3$

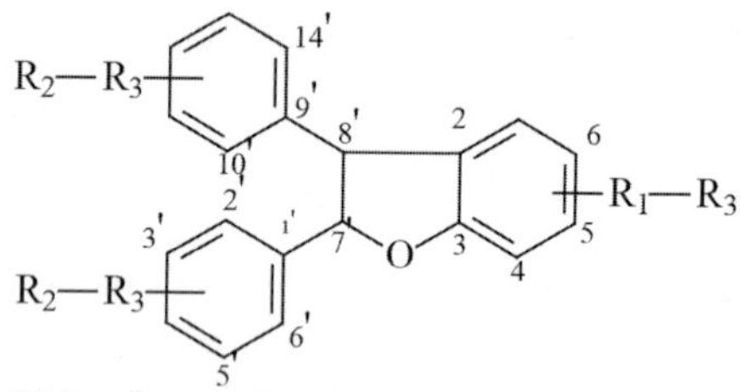

Bisisorhapontigenin

$R_1=$ 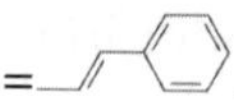 , $R_2=OH$, $R_3=OCH_3$

Experimental Materials

Samples

Rhizoma belamcandae samples were collected from Nanjing and Jiangpu of Jiangsu Province in late October 2002. There were 12 samples with variant identifications completed on all of them.

Reagents

Acetonitrile (chromatographic-grade, Tedia Company Inc., USA). Double-distilled deionized water (prepared by our lab). Methanol (analytical-grade methanol was purchased from Shanghai Chemical Reagents Pte. Ltd.; chromatographic-grade methanol was purchased from Tedia Company Inc.).

Preparation of Samples

2 g of crude powder of the sample was accurately weighed and placed in a round-bottom flask. 100 ml of methanol was added to the flask. After 2 h of extraction by reflux, the mixture was filtered. 100 ml of methanol was added to the residue and the solution was extracted for another 1 h. After filtering, the two filtrates were combined. The filtrate was evaporated to dryness on an evaporating dish. The residue was dissolved and topped up to 10 ml with methanol in a volumetric flask. The mixture was filtered through a 0.45 μm Millipore filter. The filtrate was then ready to be used as a test sample.

HPLC Conditions

Instrument: Agilent 1100 high-performance liquid chromatography and diode array detector
Mobile phase: Gradient elution with water and acetonitrile
Flow rate: 1.0 ml/min
Column: Agilent ZORBAX 80A Extend-C_{18} analytical column (250 mm × 4.6 mm, 5 μm)
Column temperature: 25°C
Detecting wavelength: 230 nm
Reference wavelength: 360 nm
Run time: 40 min
Injection volume: 10 μl

Methods

The sample solution was injected into the liquid chromatograph. The chromatogram was obtained under selected HPLC conditions. The experimental method was validated.

Mobile Phase Selection

A comparison of the experimental results showed that gradient elution with water and acetonitrile could successfully separate the chromatographic peaks and produce the maximum number of peaks from the samples. The duration of the analysis was 40 min. The 1.5 h chromatogram of the samples showed no characteristic peaks after 40 min.

A is water, B is acetonitrile.

	A (%)	B (%)
0–40 min	46–98	2–54

Method Validation

1. Stability: Test samples from the same batch were stored at room temperature and injected at regular intervals for testing. The degrees of similarity of the injections calculated by the Similarity Degree Calculation Software for Traditional Chinese Medicine Fingerprints were all above 0.93, which indicated that samples were stable within the 36 testing hours.
2. Precision: Five consecutive injections of the samples from the same batch were carried out to determine the precision. The degrees of similarity of the injections calculated by the Similarity Degree Calculation Software for Traditional Chinese Medicine Fingerprints were all above 0.98, which indicated that the injections of the test samples were carried out with precision.
3. Reproducibility: Five samples from the same batch were prepared and tested according to the method described above. The degrees of similarity of the injections calculated by the Similarity Degree

Calculation Software for Traditional Chinese Medicine Fingerprints were all above 0.94, indicating good reproducibility of the results.

Sample Analysis

The HPLC fingerprints of the test samples from 10 batches were analyzed under the conditions described above.

Establishment of Fingerprints

The number of peaks, peak area and relative retention time of 10 batches of samples were analyzed and compared to obtain an optimized fingerprint (Figs. 34.1, 34.2).

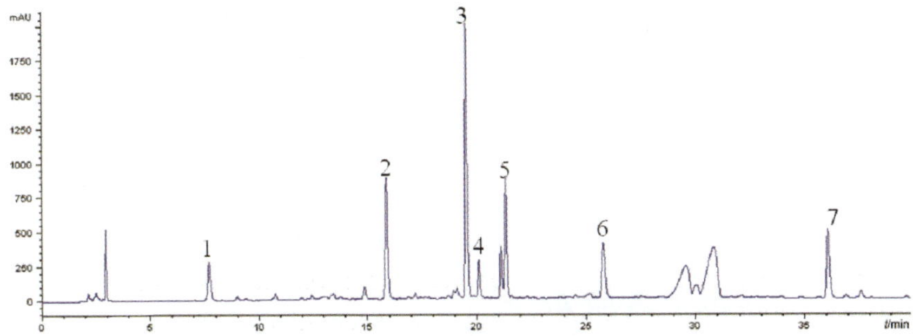

Fig. 34.1. HPLC fingerprint of Shegan (*Rhizoma belamcandae*).

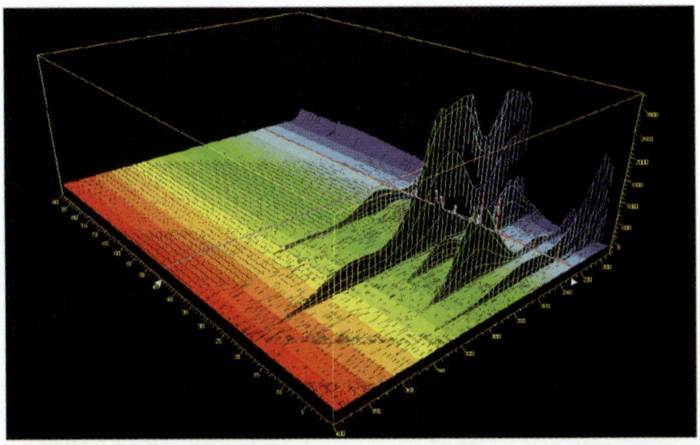

Fig. 34.2. 3-D HPLC fingerprint of Shegan (*Rhizoma belamcandae*).

Results and Discussion

About 50 peaks were separated using HPLC. After comparing all the tested and documented fingerprints, seven common peaks were selected as the characteristic peaks of the *Rhizoma belamcandae* fingerprint. The retention time (min) and peak area values of the seven common peaks are listed below.

Peak number	1	2	3	4	5	6	7
Retention time	7.703	15.873	19.523	20.098	21.324	25.791	36.076
Peak area	2784	7563.5	14025.2	2952.1	9326.2	4609.9	4960.3

Common Peak Standardization

Common peaks were assigned according to the results of 10 batches of samples. There were seven common peaks in the HPLC fingerprints of *Rhizoma belamcandae*. The common peak of peak 3 was subsequently used as the reference peak. Compared to the reference peak, the relative retention times of the other six common peaks were 0.395, 0.813, 1.029, 1.092, 1.321 and 1.848.

Ratio of Common Peak Area

Aside from the reference peak, none of the common peak areas exceeded 10% of the total peak area.

2.35

Xuchangqing
徐长卿
Radix Cynanchi paniculati

Introduction

Radix Cynanchi paniculati is the dried root and rhizome of *Cynanchum paniculatum* (Bge.) Kitag. (family Asclepiadaceae). It is mainly produced in Jiangsu, Zhejiang, Anhui, Shandong, Hubei, Hunan and Henan Provinces.

According to studies on resources, *Radix Cynanchi paniculati* is planted abundantly in Zhenjiang, Lianyungang, Xutai, Nanjing and Yixing of Jiangsu Province. The plants used for medicinal purposes are usually collected as whole plants (with roots).

Chemical Composition

The root and rhizome contain around 2% of paeonol and also isopaeonol, β-sitosterol, stearic acid, triacontane and hexadecylene. They also contain steroids and their glycosides, such as sarcostin, cynatratosides A, B, C, cynapanosides A, B, C, and neocynapanoside.

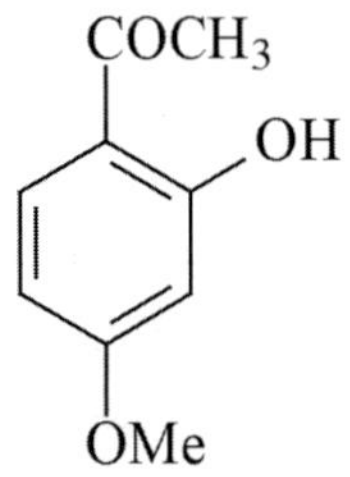

Paeonol

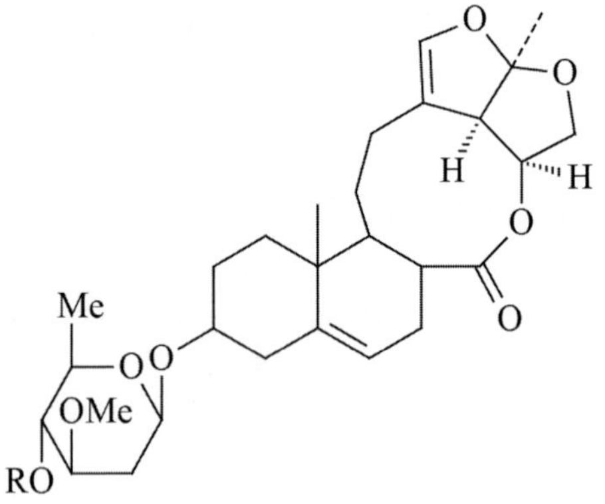

Cynapanoside A R=H
Cynapanoside B R=rha

Experimental Materials

Samples

Radix Cynanchi paniculati samples were collected from Lianyungang and Zhenjiang of Jiangsu Province. There were 12 samples with variant identifications completed on all of them.

Reference

Paeonol used for quantitative analysis was provided by the National Institute for the Control of Pharmaceutical and Biological Products (China).

Reagents

Acetonitrile (chromatographic-grade, Tedia Company Inc., USA). Acetone (analytical-grade). Methanol (analytical-grade methanol was purchased from Shanghai Chemical Reagents Pte. Ltd.; chromatographic-grade methanol was purchased from Tedia Company Inc.). Double-distilled deionized water (prepared by our lab). Chloroform (analytical-grade, Shanghai Chemical Reagents Pte. Ltd.).

Preparation of Samples

5 g of crude powder of the sample was weighed and immersed in 5 ml of methanol solution (with 1 mol/L HCl) for 2 h. The chloroform : methanol (1:1) solution was added and the solution was extracted by agitation. After filtering, the filtrate was evaporated to dryness in an evaporating dish. The residue was dissolved and topped up to 25 ml with methanol. After being filtered through a 0.45 μm Millipore filter, the filtrate was then ready to be used as a test sample.

Preparation of Reference

A proper amount of paeonol was accurately weighed and dissolved in acetone to make a 20 μg/ml solution. It was then ready to be used as a reference.

HPLC Conditions

Instrument: Agilent 1100 high-performance liquid chromatography and diode array detector
Column: Agilent ZORBAX 80A Extend-C_{18} analytical column (250 mm × 4.6 mm, 5 μm)
Mobile phase: Gradient elution with H_2O and CH_3CN
Flow rate: 1.0 ml/min
Column temperature: 25°C
Detecting wavelength: 220 nm
Reference wavelength: 360 nm
Run time: 75 min
Injection] volume: 20 μl sample, 10 μl reference

Methods

A proper amount of sample solution was injected into the liquid chromatograph. The chromatogram was obtained under selected HPLC conditions. The experimental method was validated.

Mobile Phase Selection

A comparison of the experimental results showed that gradient elution with water and acetonitrile could successfully separate the chromatographic peaks and produce the maximum number of peaks from the samples. The duration of the analysis was 75 min. The 2 h chromatogram of the samples showed no characteristic peaks after 75 min.

A is water, B is acetonitrile.

	A (%)	B (%)
0–10 min	98–98	2–2
10–45 min	40–98	2–60
45–50 min	40–40	60–60
50–65 min	0–40	60–100
65–75 min	0–0	100–100

Method Validation

1. Stability: Test samples from the same batch were stored at room temperature and injected at regular intervals for testing. The degrees of similarity calculated by the Similarity Degree Calculation Software for Traditional Chinese Medicine Fingerprints were all above 0.95, which indicated that samples were stable within the 36 testing hours.
2. Precision: Five consecutive injections of the samples from the same batch were carried out to determine the precision. The degrees of similarity calculated by the Similarity Degree Calculation Software for Traditional Chinese Medicine Fingerprints were all above 0.99, which indicated that the injections of the test samples were carried out with precision.
3. Reproducibility: Five samples from the same batch were prepared and tested according to the method described above. The degrees of similarity calculated by the Similarity Degree Calculation Software for Traditional Chinese Medicine Fingerprints were all above 0.95, indicating good reproducibility of the results.

Sample Analysis

The HPLC fingerprints of the test samples from 10 batches were analyzed under the conditions described above.

Establishment of Fingerprints

The number of peaks, peak area and relative retention time of 10 batches of samples were analyzed and compared to obtain an optimized fingerprint (Figs. 35.1–35.3).

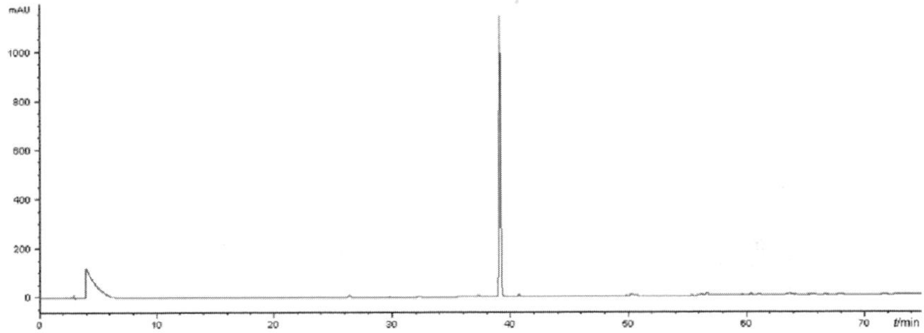

Fig. 35.1. HPLC chromatogram of paeonol.

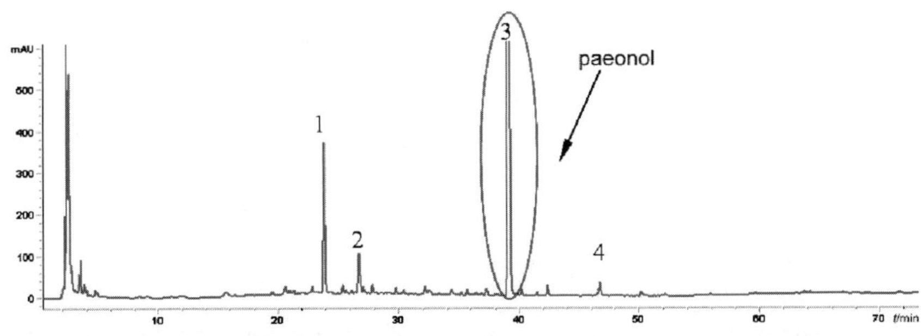

Fig. 35.2. HPLC fingerprint of Xuchangqing (*Radix Cynanchi paniculati*).

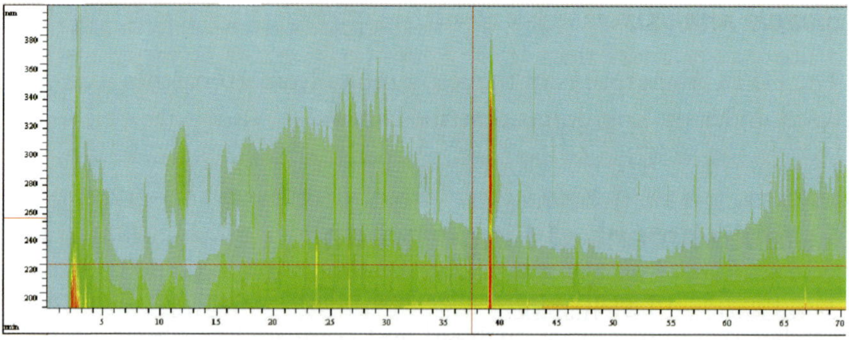

Fig. 35.3. High-resolution HPLC fingerprint of Xuchangqing (*Radix Cynanchi paniculati*).

Results and Discussion

About 20 peaks were separated using HPLC. After comparing with reference controls and comparing all the tested and documented fingerprints, six common peaks were selected as the characteristic peaks of the *Radix Cynanchi paniculati* fingerprint. The retention time (min) and peak area values of the six common peaks are listed below.

Peak number	1	2	3	4	5	6
Retention time	23.778	26.672	27.816	39.07	42.352	46.634
Peak area	3840.5	1637.1	627.5	28985	328.9	364.1

Common Peak Standardization

Common peaks were assigned according to the results of 10 batches of samples. There were six common peaks in the HPLC fingerprints of *Radix Cynanchi paniculati*. Compared to the paeonol reference peak (peak 4), the relative retention times of the other five common peaks were 0.609, 0.683, 0.712, 1.084 and 1.194.

Ratio of Common Peak Areas

Only the peak area of the paeonol reference peak (peak 4) was more than 10% of the total peak area. Aside from that, no other common peak area was more than 10% of total peak area.

Evaluation of Similarity Degrees

Experimental data were imported into the Similarity Degree Calculation Software for Traditional Chinese Medicine Fingerprints. After selecting peaks and setting a matching template, the chromatographic peaks were automatically matched. Then the standard template was determined, and was used for evaluations of degrees of difference and overall degrees of similarity of fingerprint peaks.

The common pattern of the HPLC fingerprint of *Radix Cynanchi paniculati* was obtained using the Similarity Degree Calculation Software for Traditional Chinese Medicine Fingerprints. With reference to the common pattern, degrees of similarity of samples collected from four places in Jiangsu Province were 0.95 (Lianyungang), 0.98 (Zhenjiang), 0.90 (Xutai) and 0.92 (Nanjing).

2.36

Huangqin
黄芩
Radix scutellariae

Introduction

Radix scutellariae is the dried root of *Scutellaria baicalensis* Georgi (family Labiatae). It is mainly produced in the Northeast of China, in Hebei and Shaanxi Provinces. Shaanxi Province has the highest yield of *Radix scutellariae*.

According to studies on resources, *Radix scutellariae* is planted abundantly in Chengde, Rehe, Baoding, Longhua, Fengning, Pingquan, Xinglong, Luanping, Qinglong, Kuancheng, Huai'an, Wanquan, Huailai, Guyuan, Weixian, Chicheng, Zhangbei, Jinxing, Lingshou, Pingshan, Huolu, Wu'an, Yixian, Laiyuan and Buping of Hebei Province. The quality of *Radix scutellariae* produced in Chengde is the best and the *Radix scutellariae* produced in Rehe is the most famous one.

Chemical Composition

Radix scutellariae contains various derivatives of flavones, such as baicalin, wogonoside, oroxin, baicalein, wogonin, oroxylin A, norbaicalein, skullcap flavones I, II, 7-methoxybaicalein and 7-methoxynorwogonin.

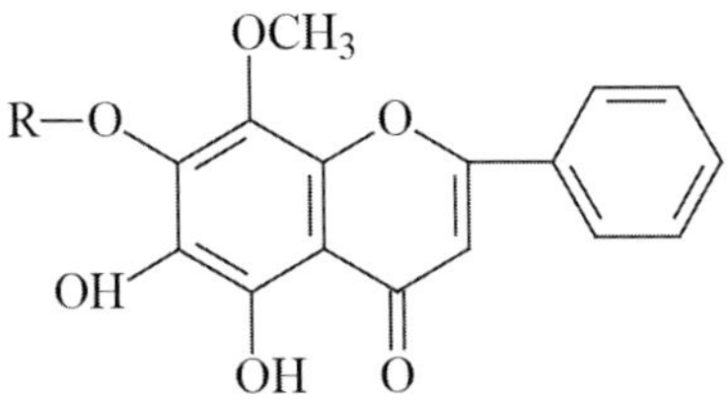

Wogonin R=H

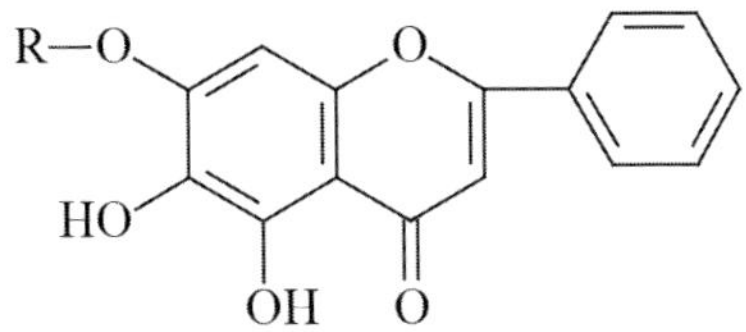

Baicalein R=H
7-Methoxybaicalein R=CH_3

Wogonoside R= 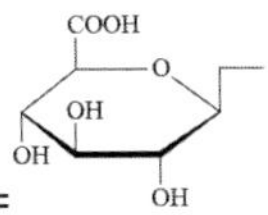

Baicalin R= 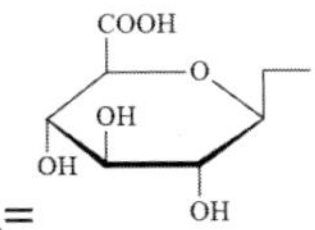

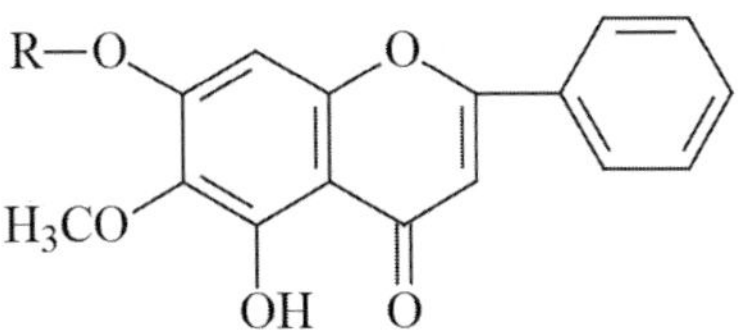

Oroxylin A R=H

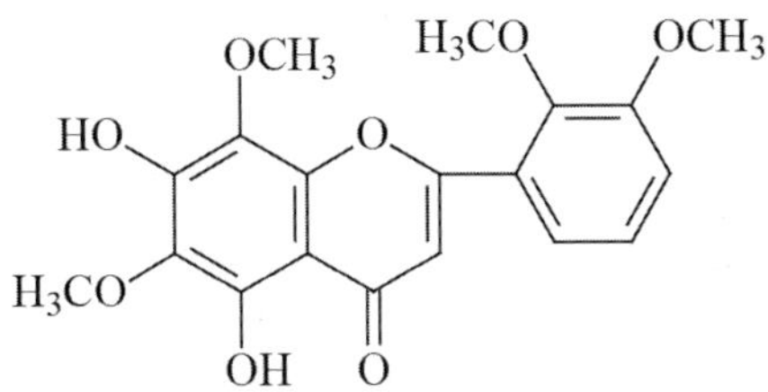

Norbaicalein

Oroxin R= 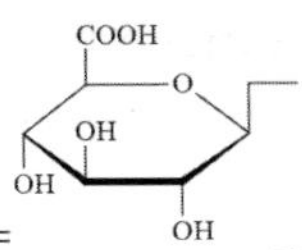

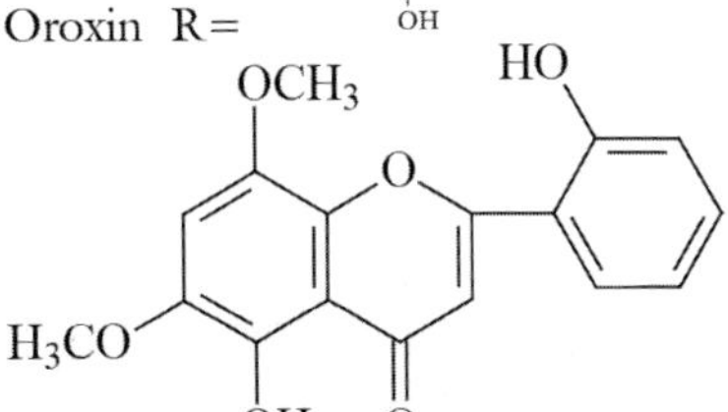

Skullcap flavone I

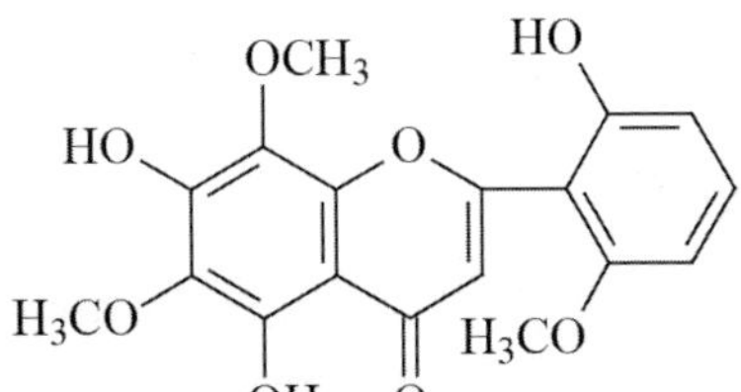

Skullcap flavone II

Experimental Materials

Samples

Radix scutellariae samples were collected from Chengde, Anguo, and Weichang of Hebei Province, Puning of Guangdong Province, Ankang, Xi'an, and Baishan of Shaanxi Province, Longxi of Gansu Province, Yulin of Guangxi Province, and Inner Mongolia Province. The samples were three-year-old or four-year-old roots. There were 13 samples with variant identifications completed on all of them.

Reference

Baicalin was provided by the National Institute for the Control of Pharmaceutical and Biological Products (China) and the purity was over 98%.

Wogonoside was prepared by our lab and its structure was confirmed by analyzing the NMR and FAB-MS data and comparing it with the documented NMR data of wogonin. Its purity calculated by HPLC normalization was over 99%.

Baicalein control was provided by Yoneyama Yakuhin Kogyo Co. Ltd., Japan. Its purity calculated by HPLC normalization was over 96%.

Wogonin control was prepared by our lab and its structure was confirmed by analyzing the NMR and MS data and referring to related documentations. Its purity calculated by HPLC normalization was over 98%.

Reagents

Acetonitrile (CH_3CN) (chromatographic-grade). Water (Milli-Q H_2O). Tetrahydrofuran (THF), phosphoric acid (H_3PO_4), methanol and 70% ethanol were all analytical-grade products.

Preparation of Samples

0.2 g of powder sample (passed through a 3 mesh sieve) was accurately weighed and placed in a 100 ml volumetric flask. It was topped up to the mark with 70% ethanol and mixed well. The solution was sonicated for 30 min, allowed to cool, and was filtered. 2 ml of filtrate was placed in a 5 ml volumetric flask and was topped up to the mark with 70% ethanol. The solution was mixed well and filtered through a 0.45 μm Millipore filter. The filtrate was then ready to be used as a test sample.

Preparation of Reference

10 ml of baicalin (0.250 mg/ml), 7 ml of wogonoside (0.0868 mg/ml), 1.5 ml of baicalein (0.106 mg/ml) and 0.5 ml of wogonin (0.148 mg/ml) were placed in the same 25 ml volumetric flask and topped up to the mark with methanol. After mixing well, it was ready to be used as a reference.

HPLC Conditions

Instrument: Waters high-performance liquid chromatography, LC-2690 pump, 996 diode array detector, 600 autosampler, Millennium 32 data analysis software system.
Mobile phase: Gradient elution with CH_3CN and H_2O-THF-H_3PO_4 (80:10:0.2)
Flow rate: 1.0 ml/min
Column: Platinum (Alltech) C_{18} (4.6 mm × 250 mm, 5 μm)
Column temperature: 25°C
Detecting wavelength: 274 nm
Run time: 60 min
Injection volume: 10 μl sample, 10 μl reference

Methods

A proper amount of sample solution was injected into the liquid chromatograph. The chromatogram was obtained under selected HPLC conditions. The experimental method was validated.

Mobile Phase Selection

A comparison of the experimental results showed that gradient elution with CH_3CN and H_2O-THF-H_3PO_4 (80:10:0.2) could successfully separate the chromatographic peaks and produce the maximum number of peaks from the samples. The duration of the analysis was 60 min. The 1.5 h chromatogram of the samples showed no characteristic peaks after 60 min.

A is CH_3CN, B is H_2O-THF-H_3PO_4 (80:10:0.2).

	A (%)	B (%)
0–15 min	20–20	80–80
15–45 min	20–30	70–80
45–55 min	30–30	70–70
55–60 min	20–30	70–80

Method Validation

1. Precision: Six consecutive injections of the samples were conducted to determine the precision. For every injection, 10 μl of reference solution was accurately aspirated. The peak areas were measured. The RSDs of baicalin, wogonoside, baicalein and wogonin were 0.74%, 1.1%, 1.1% and 1.6%, respectively, which indicated the injections of the samples were carried out with precision.
2. Reproducibility: Six samples from the same batch were prepared and tested according to the HPLC conditions described above. The average concentrations (n = 6) of baicalin, wogonoside, baicalein and wogonin were 11.34%, 2.92%, 0.80% and 0.26%, respectively. The RSDs were 1.0%, 1.5%, 2.6% and 3.1%, respectively, indicating good reproducibility of the results.
3. Sample recovery rate: 0.15 g of the sample with known concentra-

tion was accurately weighed. The following reference solutions were added: 4 ml of baicalin (0.807 mg/ml), 2 ml of wogonoside (0.488 mg/ml), 3 ml of baicalein (0.141 mg/ml) and wogonin (0.155 mg/ml). The samples were prepared according to the methods mentioned above and were tested five times. The average recovery (n = 5) of baicalin, wogonoside, baicalein and wogonin were 101.5%, 99.5%, 98.9% and 100.4%, respectively. The RSDs were 2.2%, 2.3%, 1.0% and 3.0%, respectively.

Sample Analysis

The HPLC fingerprints of the test samples from 10 batches were analyzed under the conditions described above.

Establishment of Fingerprints

The number of peaks, peak area and relative retention time of 10 batches of samples were analyzed and compared to obtain an optimized fingerprint (Fig. 36.1).

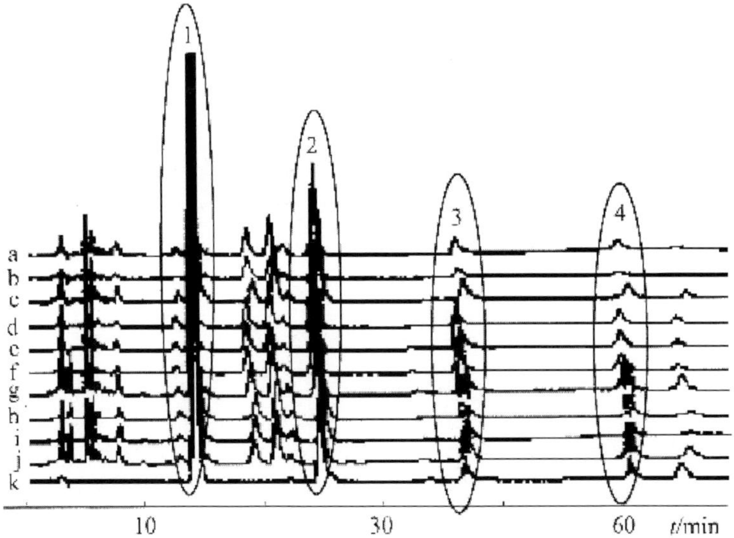

Fig. 36.1. HPLC fingerprint of Huangqin (*Radix scutellariae*).
Note: a–j: test samples; k: reference; 1: baicalin; 2: wogonoside; 3: baicalein; 4: wogonin.

Results and Discussion

About 20 peaks were separated using HPLC. After comparing with the reference controls and comparing all the tested and documented fingerprints, 14 common peaks were selected as the characteristic peaks of the *Radix scutellariae* fingerprint.

Common Peak Standardization

Common peaks were assigned according to the results of 10 batches of samples. There were 14 common peaks in the HPLC fingerprints of *Radix scutellariae*. Compared to the baicalin reference peak (peak 1), the relative retention times of the other 13 common peaks were 0.37, 0.41, 0.45, 0.48, 0.56, 0.91, 1.34, 1.47, 1.56, 1.74, 2.59, 3.57 and 3.90.

Ratio of Common Peak Areas

Aside from the baicalin reference peak, none of the common peak areas exceeded 10% of the total peak area.

2.37

Huangqi
黄芪
Radix astragali

Introduction

Radix astragali (family Leguminosae) is the dried root of *Astragalus membranaceus* (Fisch.) Bge. var. *mongholicus* (Bge.) Hsiao, also called Menggu Huangqi (蒙古黄芪); or the dried root of *Astragalus membranaceus* (Fisch.) Bge., also called Mojia Huangqi (膜荚黄芪).

Menggu Huangqi is mainly produced in Shanxi, Inner Mongolia and Hebei, while Mojia Huangqi is produced in Heilongjiang, Jilin, Inner Mongolia and Hebei.

According to studies on resources, it has been found that Menggu Huangqi is grown in abundance in Hunyuan, Yingxian, Yanggao, Tianzhen, Daixian, Fanzhi, Wutai, etc, of Yanbei in Shanxi Province, and its quality is extremely good. The cultivation area is 16,000 hm^2 in Hunyuan and Yingxian. Sixty percent of exported *Radix astragali* is from these two places. It is also produced in Pinglu, Pingshun, Ningwu, Loufan, Kelan, Suxian, Jiexiu, Jiaocheng, etc.

Chemical Composition

Radix astragali mainly contains saponins, flavonoids and polysaccharides. Saponin components include astragalosides I, II and IV, acetylastragaloside I, isoastragalosides I and II. Flavonoid components include 2′,4′-dihydroxy-5,6-dimethoxyisoflavane. It also contains γ-aminobutyric acid.

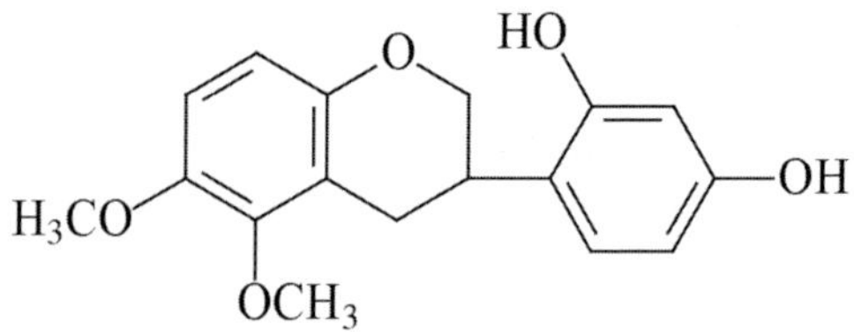

2′, 4′-Dihydroxy-5,6-dimethoxyisoflavane

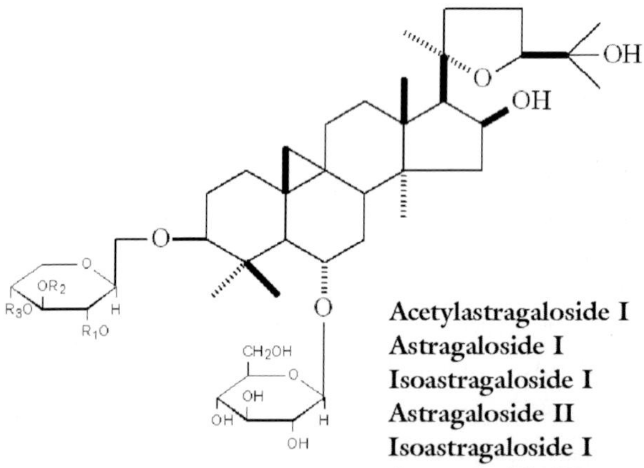

	R_1	R_2	R_3
Acetylastragaloside I	Ac	Ac	Ac
Astragaloside I	Ac	Ac	H
Isoastragaloside I	Ac	H	Ac
Astragaloside II	Ac	H	H
Isoastragaloside I	H	Ac	H
Astragaloside IV	H	H	H

Experimental Materials

Samples

Radix astragali samples were collected from three-year-old plants in Hunyuan and Yingxian of Yanbei in Shanxi Province in October 2002. There were 15 batches of samples collected and identified as *Astragalus membranaceus*.

Reference

Ginsenoside Re used for quantitative analysis was provided by the National Institute for the Control of Pharmaceutical and Biological Products (China).

Reagents

Acetonitrile (chromatographic-grade, Tedia Company Inc., USA). Double-distilled deionized water (prepared by our lab). Methanol (analytical-grade methanol was purchased from Shanghai Chemical Reagents Pte. Ltd.; chromatographic-grade methanol was purchased from Tedia Company Inc.).

Preparation of Samples

2 g of crude powder was weighed and 2 x 50 ml of methanol was used for Soxhlet extraction for 2 h. The methanol was collected. The residue was dissolved in 10 ml of warm water and extracted five times with an equal amount of water-saturated *n*-butanol. The *n*-butanol was combined and washed once with an equal amount of ammonia solution. The ammonia solution was discarded and *n*-butanol was evaporated to dryness over a water bath. The residue was dissolved and topped up to 5 ml by adding 60% acetonitrile. The mixture was centrifuged and filtered. 1 ml of filtrate was collected and 1 ml of internal reference was added and mixed well for analysis.

Preparation of Reference

An appropriate amount of ginsenoside Re was accurately weighed and methanol was added to obtain a concentration of 0.1 mg/ml. It was then ready to be used as an internal reference.

HPLC Conditions

Instrument: Agilent 1100 high-performance liquid chromatography and variable wavelength detector
Mobile phase: Gradient elution with water and acetonitrile
Flow rate: 1.0 ml/min
Column: Hypersil ODS (250 mm × 4.6 mm, 5 μm)
Column temperature: 30°C

Detecting wavelength: 203 nm
Run time: 60 min
Injection volume: 20 μl sample, 10 μl reference

Methods

The sample was injected into the liquid chromatograph. The chromatogram was obtained under selected HPLC conditions. The experimental method was validated.

Mobile Phase Selection

A comparison of experimental results showed that gradient elution with water and acetonitrile could successfully separate the peaks and produce the maximum number of peaks from the samples. The duration of the analysis was 60 min. The 1.5 h chromatogram of the samples showed no characteristic peaks after 60 min.

A is H_2O, B is acetonitrile.

	A (%)	B (%)
0–50 min	50–100	0–50
50–60 min	50–50	50–50

Method Validation

1. Stability: The sample solution from the same batch was injected and analyzed at 0 h, 2 h, 4 h, 8 h, 16 h, 24 h and 48 h. The RSD of the relative peak area was 2%, which indicated good stability of the sample solution.
2. Precision: Five consecutive injections of the samples from the same batch were conducted. The RSD of the relative peak area was 1%, which meets our criteria.
3. Reproducibility: Five samples from the same batch were prepared using the method described above and tested simultaneously to determine reproducibility. The RSD of the relative peak area was less than 3%, which meets our criteria.

Sample Analysis

Ten batches of samples were analyzed under the conditions described above.

Establishment of Fingerprints

The number of peaks, peak area and relative retention time of the samples from the 10 batches were analyzed and compared to establish an optimized fingerprint (Fig. 37.1).

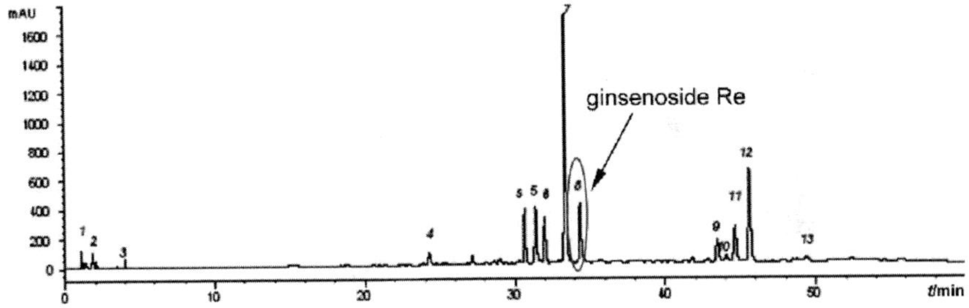

Fig. 37.1. HPLC fingerprint of Huangqi (*Radix astragali*).

Results and Discussion

Common Peak Standardization

Common peaks were assigned according to the results of 10 batches of samples. There were 13 common peaks in the HPLC fingerprint of *Radix astragali*. The peak of ginsenoside Re (internal standard) was assigned as the reference peak, and the relative retention times of the other common peaks were 0.04, 0.06, 0.13, 0.79, 1.03, 1.05, 1.09, 1.13, 1.42, 1.44, 1.46, 1.49 and 1.62.

Characteristics of Common Peaks

Peaks 1–3 were multiple peaks, while peaks 4 and 5 were multiple jagged peaks.

Ratio of Common Peak Area

The peak areas of common peaks 7 and 12 exceeded 10% of the total peak area. With the peak area of common peak 12 designated as 1, the relative peak area of common peak 7 was specified to be within 1.141–2.119.

2.38

Huanglian

黄连

Rhizoma coptidis

Introduction

Rhizoma coptidis (family Ranunculaceae) is the dried rhizome of *Coptis chinensis* Franch., *Coptis deltoidea* C.Y. Cheng et Hsiao, or *Coptis teeta* Wall. They are commonly referred to as Weilian, Yalian and Yunlian, respectively. Among them, Weilian is the main species of Chinese medicinal *Rhizoma coptidis*.

It is mainly produced in Sichuan and Yunnan Provinces and is in abundance in Sichuan. Weilian is mainly found in Shizhu and Lichuan in Sichuan Province, while Yunlian is mainly produced in Deqin and Bijiang of Yunnan Province and Tibet. According to studies on resources, Yalian is cultivated in abundance in Hongya, Emei, Ebian, Jinkouhe, Mabian, Leshan and Leibo of Sichuan Province. It is renowned for its abundance in Hongya and Emei.

Chemical Composition

All three kinds of *Rhizoma coptidis* contain many kinds of isoquinoline alkaloids, mainly berberine, and in lesser amounts, coptisine, worenine, palmatine and jatrorrhizine. It also contains magnoflorine and epiberberine.

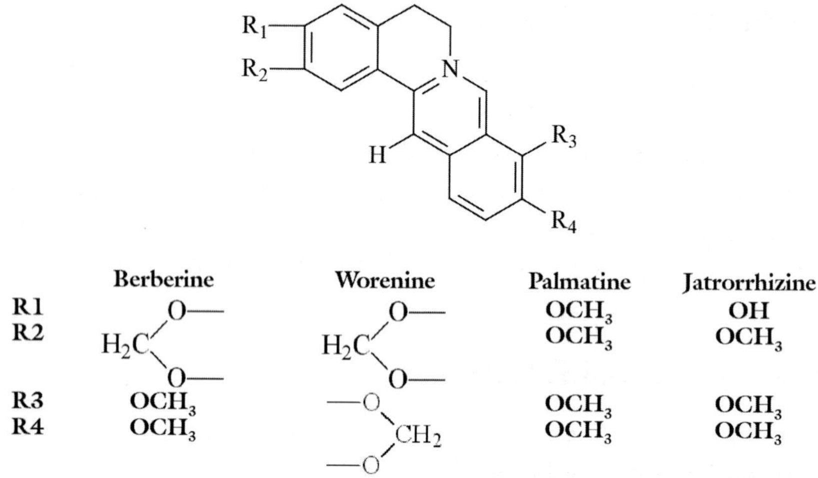

Experimental Materials

Samples

Rhizoma coptidis samples were five-year-old rhizomes collected from Hongya and Emei, etc, in Sichuan Province in November 2001. There were 12 batches of samples identified as *Coptis deltoidea* C.Y. Cheng et Hsiao (Yalian).

Reference

Berberine was prepared by our laboratory.

Reagents

Acetonitrile (chromatographic-grade, Tedia Company Inc., USA). Methanol (analytical-grade, Shanghai Chemical Reagent Pte. Ltd.). Hydrochloric acid and potassium dihydrogen phosphate (analytical-grade, First Nanjing Chemical Reagent Company). Sodium hydroxide (analytical-grade, Shanghai Chemical Reagent Pte. Ltd.). Double-distilled deionized water (prepared by our lab).

Preparation of Samples

0.1 g of crude powder was accurately weighed and placed in a 100 ml volumetric flask. About 95 ml of methanol with 1% HCl was added and the mixture was heated in a 60°C water bath for 15 min. It was then sonicated for 30 min (operating frequency of 40 kHz; power of 500 W) and stored at room temperature overnight. Thereafter it was topped up to volume with methanol, mixed well and filtered.

Preparation of Reference

Berberine standard was weighed accurately and dissolved in methanol to make a 40 μg/ml solution.

HPLC Conditions

Instrument: Agilent 1100 high-performance liquid chromatography and diode array detector
Mobile phase: Phosphate buffer solution and acetonitrile with gradient elution
Flow rate: 1.0 ml/min
Column: Agilent ZORBAX 80A Extend-C_{18} analytical column (250 mm × 4.6 mm, 5 μm)
Column temperature: 25°C
Detecting wavelength: 230 nm
Reference wavelength: 360 nm
Run time: 65 min
Injection volume: 10 μl

Methods

A proper amount of sample solution was injected into the liquid chromatograph. The chromatogram was obtained under selected HPLC conditions. The experimental method was validated.

Mobile Phase Selection

A comparison study showed that better separation in the chromatograph could be achieved with gradient elution using phosphate buffer and acetonitrile, and the maximum number of peaks could be successfully produced. The duration of analysis was 65 min. The 2 h chromatogram showed no characteristic peaks after 65 min.

A is phosphate buffer and B is acetonitrile.

	A (%)	B (%)
0–25 min	70–98	2–30
25–45 min	55–70	30–45
45–60 min	10–55	45–90
60–65 min	10–10	90–90

Method Validation

1. Stability: The sample solution from the same batch was stored at room temperature and injected at regular intervals. The degrees of similarity of all injections calculated by the Similarity Degree Calculation Software for Traditional Chinese Medicine Fingerprints were all above 0.96, which indicated that the sample was stable within the 36 testing hours.
2. Precision: Five consecutive injections of the samples from the same batch were analyzed to determine the precision. The degrees of similarity calculated by the Similarity Degree Calculation Software for Traditional Chinese Medicine Fingerprints were all above 0.97, which indicated that the injection method was carried out with acceptable precision.
3. Reproducibility: Five samples from the same batch were prepared with the method described above and tested individually to determine reproducibility. The degrees of similarity of the five samples calculated by the Similarity Degree Calculation Software for Traditional Chinese Medicine Fingerprints were all above 0.95, indicating that the sample preparation method was reproducible.

Sample Analysis

Ten batches of samples were analyzed under the conditions described above.

Establishment of Fingerprints

The number of peaks, peak area and relative retention time of the samples from the 10 batches were analyzed and compared to establish an optimized fingerprint (Figs. 38.1 and 38.2).

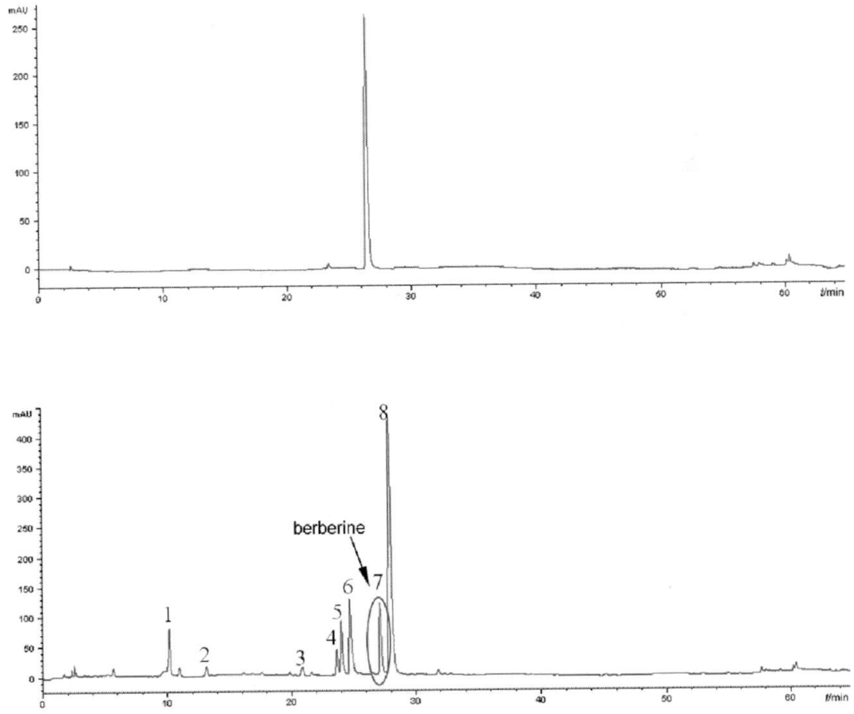

Fig. 38.1. HPLC fingerprints of berberine (top) and of Huanglian (*Rhizoma coptidis*) (bottom).

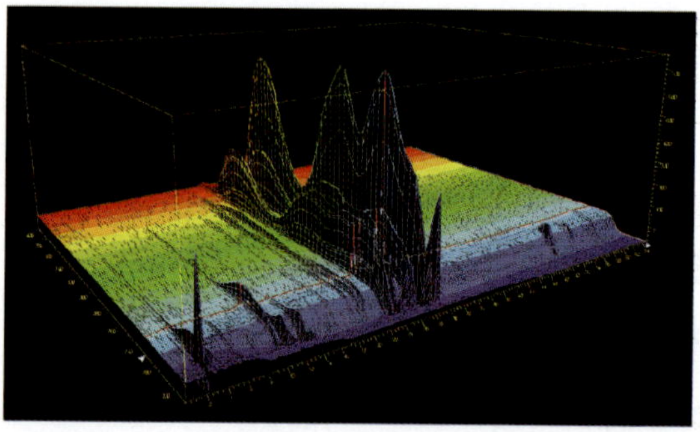

Fig. 38.2. 3-D HPLC fingerprint of Huanglian (*Rhizoma coptidis*).

Results and Discussion

Over 20 peaks were separated under the above HPLC conditions. Eight common peaks were selected to characterize the fingerprint of *Rhizoma coptidis* by comparing them with all the documented fingerprints and with the reference standard. The retention time (min) and peak area values are listed below.

Peak number	1	2	3	4	5	6	7	8
Retention time	10.126	13.084	20.825	23.624	23.987	24.677	27.105	27.859
Peak area	1041	190.8	200.8	410.9	1081.9	1844.4	1390.8	7681.2

Common Peak Standardization

Common peaks were assigned according to the results of 10 batches of samples. There were eight common peaks in the HPLC fingerprint of *Rhizoma coptidis*. Compared to the reference peak of berberine (peak 7), the relative retention times of the other seven common peaks were 0.374, 0.483, 0.768, 0.872, 0.885, 0.910 and 1.028.

Ratio of Common Peak Area

The peak areas of common peaks 6 and 8 exceeded 10% of the total peak area. The relative peak area ratio of peak 6 was specified to be within 0.9283–1.7240, whilst that for peak number 8 was within 3.8660–7.1797.

2.39

Yinxingye
银杏叶
Folium ginkgo

Introduction

Folium ginkgo is the dried leaf of *Ginkgo biloba* L. (family Ginkgoaceae). It is mainly cultivated in Guangxi, Sichuan, Hubei, Zhejiang, Jiangsu and Shandong Provinces.

According to studies on resources, *Folium ginkgo* is planted abundantly in some areas of Jiangsu Province, such as Taizhou, Suzhou and Pizhou.

Chemical Composition

Folium ginkgo mainly contains ginkgolides, such as ginkgolides A, B, C and M, and biflavonoids, such as bilobetin, ginkgetin, and sciadopitisin. It also contains quercetin, isorhamnetin, kaempferol, β-sitosterol and shikimic acid. Some of the chemical structures are listed below.

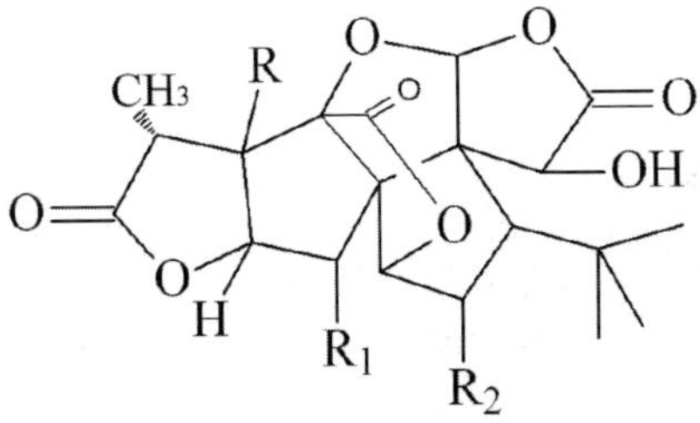

	R	R_1	R_2
Ginkgolide A	**OH**	**H**	**H**
Ginkgolide B	**OH**	**OH**	**H**
Ginkgolide C	**OH**	**OH**	**OH**

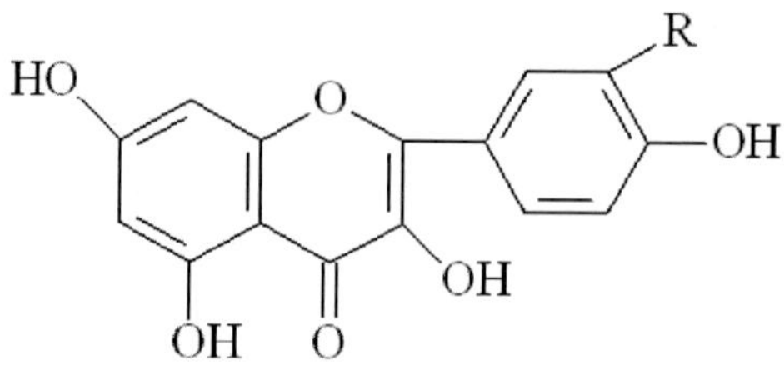

	R
Rhamnetin	OH
Isorhamnetin	OCH_3
Kaempferol	H

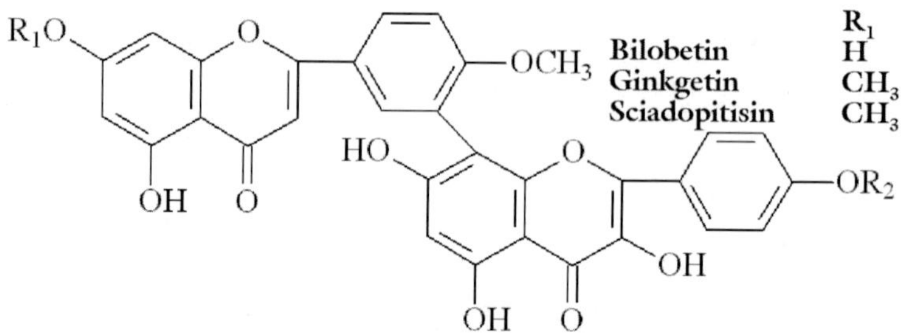

	R_1	R_2
Bilobetin	H	H
Ginkgetin	CH_3	H
Sciadopitisin	CH_3	CH_3

Experimental Materials

Samples

There were 12 batches of samples, collected from Pizhou, Jiangsu Province, and identified as *Folium ginkgo*.

References

Bilobalide and ginkgolides A and B used for quantitative analysis were provided by the National Institute for the Control of Pharmaceutical and Biological Products (China).

Reagents

Methanol (analytical-grade methanol was purchased from Shanghai Chemical Reagents Pte. Ltd.; chromatographic-grade methanol was purchased from Tedia Company Inc.). Double-distilled deionized water (provided by our lab).

Preparation of Samples

10 g of powdered sample (passed through a 20 mesh sieve) was accurately weighed. 100 ml of 60% ethanol was added and the mixture was kept warm in a 70°C water bath with constant stirring for 5 h. The mixture was filtered, and residue was extracted in the same way for 3 h. Both filtrates were combined and dried under vacuum until there was no smell of ethanol. The concentrate was diluted to 20 ml with H_2O and the solution was passed through a pre-treated D101 macroporous resin column for 30 min to let the absorption occur. The column was eluted with water until the eluate was colorless. After eluting with 95% ethanol, 500 ml of eluate was collected and concentrated under vacuum to a small volume. Thereafter 1 g of polyamide (passing through a 80 mesh sieve) was mixed with the concentrate; the mixture was dried and loaded into a polyamide column (80 mesh sieve, 4 g, diameter 2.5 cm). After eluting with 30% ethanol, 200 ml of the eluate was collected and the ethanol was evaporated off at low heat. Sodium chloride was added to the concentrated solution until saturated. The saturated solution was extracted three times with 20 ml, 20 ml and 10 ml of ethyl acetate respectively. The ethyl acetate was combined and evaporated off. The residue was dissolved and topped up to 10 ml with acetone:H_2O (1:1). The mixture was mixed well and filtered through a 0.45 μm membrane filter. The first 5 ml of the filtrate was discarded and the remaining filtrate was collected. 200 μl of the filtrate was accurately transferred and topped up to 1 ml with acetone:H_2O (1:1).

Preparation of Reference

An appropriate amount of bilobetin, ginkgolides A and B were accurately weighed and made into a 0.125 mg/ml solution by adding methanol. It was then ready to be used as a reference.

HPLC Conditions

Instrument: Agilent 1100 high-performance liquid chromatography

and diode array detector
Mobile phase: Double-distilled water and methanol with gradient elution
Flow rate: 1.0 ml/min
Column: Hypersil ODS 3 analytical column (250 mm × 4 mm, 5 μm)
Column temperature: 30°C
Detecting wavelength: 220 nm
Reference wavelength: 360 nm
Run time: 60 min
Injection volume: 10 μl sample, 20 μl reference

Methods

The sample solution was injected into the liquid chromatograph. The chromatogram was obtained under selected HPLC conditions. The experimental method was validated.

Mobile Phase Selection

A comparison study showed that better separation in the chromatograph could be achieved with gradient elution using water and methanol, and the maximum number of peaks could be successfully produced. The duration of analysis was 30 min. The 1.5 h chromatogram showed no characteristic peaks after 60 min.

A is double-distilled water and B is methanol.

	A (%)	B (%)
0–60 min	75–85	15–25
60–75 min	0–75	25–100
75–80 min	0–85	15–100

Method Validation

1. Stability: Samples from the same batch were stored at room temperature and injected at regular intervals. The degrees of similarity of the injections calculated by the Similarity Degree Calculation Software for Traditional Chinese Medicine Fingerprints were all above 0.95, which indicated that the samples were stable within the 36 testing hours.
2. Precision: Five consecutive injections of the samples from the same batch were conducted to determine the precision. The degrees of similarity of the injections calculated by the Similarity Degree Calculation Software for Traditional Chinese Medicine Fingerprints were all above 0.9, which indicated that injections of the test samples were carried out with acceptable precision.
3. Reproducibility: Five samples from the same batch were prepared and tested according to the method described above. The degrees of similarity of the injections calculated by the Similarity Degree Calculation Software for Traditional Chinese Medicine Fingerprints were all above 0.95, indicating good reproducibility of the results.

Sample Analysis

The HPLC fingerprints of the test samples from 10 batches were analyzed under the conditions described above.

Establishment of Fingerprints

The number of peaks, peak area and relative retention time of 10 batches of samples were analyzed and compared to obtain an optimized fingerprint (Figs. 39.1, 39.2).

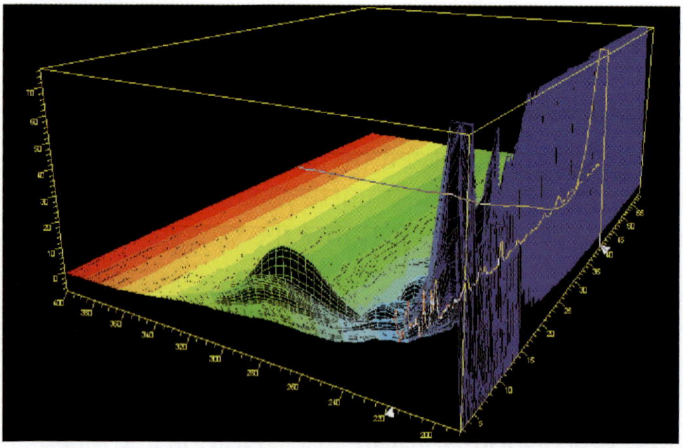

Fig. 39.1. HPLC fingerprint of Yinxingye (*Folium ginkgo*).

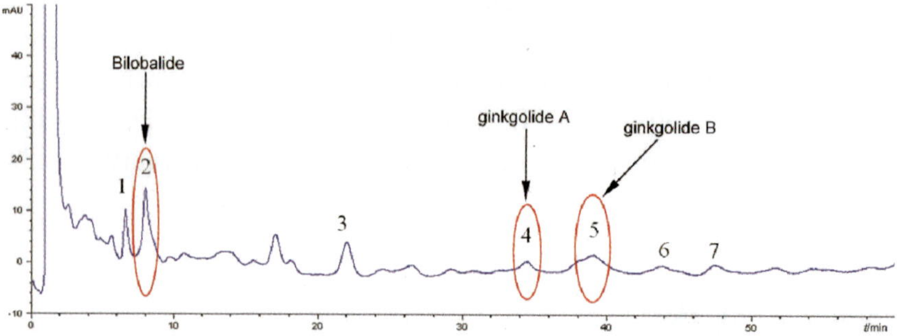

Fig. 39.2. 3-D HPLC fingerprint of Yinxingye (*Folium ginkgo*).

Results and Discussion

Over 30 peaks were separated using HPLC. After comparing all the chromatograms obtained, seven common peaks were selected to be the characteristic peaks of the *Folium ginkgo* fingerprint. The retention time (min) and peak area values of the seven common peaks are listed below.

Peak number	1	2	3	4	5	6	7
Retention time	6.6	8.07	22.06	34.53	39	43.83	47.4
Peak area	236	615.33	338.2	219.55	568.06	167.52	158.55

Common Peak Standardization

Common peaks were assigned according to the results of 10 batches of samples. There were seven common peaks in the HPLC fingerprints of *Folium ginkgo*. Compared to the reference peak of bilobetin (peak 2), the relative retention times of the other six common peaks were 0.818, 2.734, 4.279, 4.833, 5.431 and 5.874.

2.40

Yinchaihu
银柴胡
Radix stellariae

Introduction

Radix stellariae is the dried root of *Stellaria dichotoma* L. var. *lanceolata* Bge. *Radix stellariae* is mainly found in the provinces of Ningxia, Gansu and Inner Mongolia. According to studies on resources, good-quality *Radix stellariae* is cultivated in abundance in Taole, Yanchi, Lingwu, Tongxin, Zhongning, Pingluo, Helan and Yinchuan of Ningxia Province. It is planted in Guyuan, Pengyang and Longde of Ningxia Province as well.

Chemical Composition

Radix stellariae contains furoic acid, 6,8-di-C-galactopyranosyl apigenin, wogonin, wild 6-C-galactopyranosyl-isoscutellarein and stigmasterols, spinasterols and volatile saponins. A new cyclic hexapeptide is also isolated from *Radix stellariae* and is called stellaria cyclopeptide.

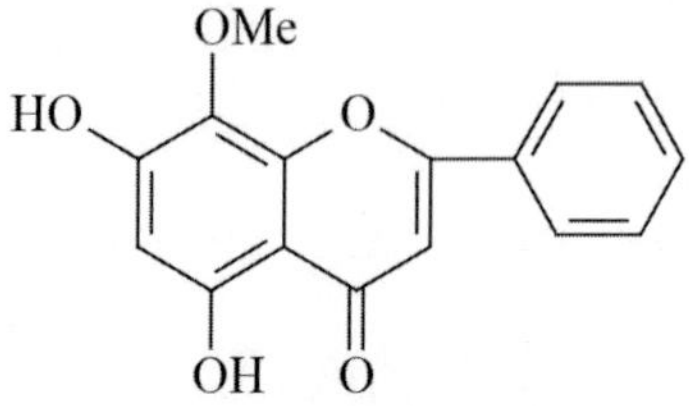

Wogonin

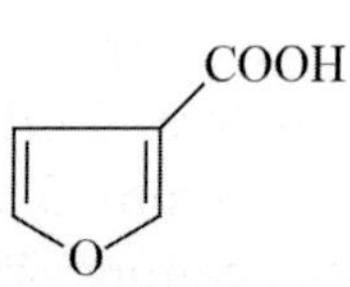

Furan-3-carboxylic acid

Experimental Materials

Samples

Radix stellariae samples were collected from three-year-old plants in Yinchuan and Pengyang in Ningxia in October 2002. There were 12 batches of the sample with variant identification completed on them.

Reagents

Acetonitrile (chromatographic-grade, Tedia Company Inc., USA). Methanol (analytical-grade methanol was purchased from Shanghai Chemical Reagents Pte. Ltd.; chromatographic-grade methanol was purchased from Tedia Company Inc.). Ether (analytical-grade, Shanghai Chemical Reagents Pte. Ltd.). Double-distilled deionized water (prepared by our lab).

Preparation of Samples

10 g of powder sample was accurately weighed and Soxhlet extracted with 100 ml of methanol for 24 h. The extract was concentrated to 10 ml and 5 ml of distilled water was added. The mixture was extracted 10 times with ether. The ether was combined and evaporated to dryness. The residue was dissolved in methanol and made up to 25 ml. The solution was filtered through a 0.45 μm Millipore filter for analysis.

HPLC Conditions

Instrument: Agilent 1100 high-performance liquid chromatography and diode array detector
Mobile phase: H_2O and acetonitrile with gradient elution
Flow rate: 1.0 ml/min
Column: Agilent ZORBAX 80A Extend-C_{18} analytical column (250 mm × 4.6 mm, 5 μm)
Column temperature: 25°C
Detecting wavelength: 210 nm

Reference wavelength: 360 nm
Run time: 70 min
Injection volume: 10 μl

Methods

A proper amount of sample solution was injected into the liquid chromatograph. The chromatogram was obtained under selected HPLC conditions. The experimental method was validated.

Mobile Phase Selection

Better separation in the chromatograph was achieved with gradient elution of water and acetonitrile after the comparison study, and more peaks were obtained. The duration of analysis was 70 min. No characteristic peaks appeared after 70 min from the extended analysis of up to 2 h.

A is water and B is acetonitrile.

	A (%)	B (%)
0–20 min	70–95	30–5
20–50 min	20–70	80–20
50–70 min	0–20	100–80

Method Validation

1. Stability: Sample solutions from the same batch were stored at room temperature and injected at regular intervals. The degrees of similarity of all injections calculated by the Similarity Degree Calculation Software for Traditional Chinese Medicine Fingerprints were all above 0.95, which indicated that the samples were stable within the 36 testing hours.
2. Precision: Five consecutive injections of the samples from the same batch were analyzed to determine the precision. The degrees

of similarity calculated by the Similarity Degree Calculation Software for Traditional Chinese Medicine Fingerprints were all above 0.94, which indicated that the injection method was carried out with acceptable precision.

3. Reproducibility: Five samples from the same batch were prepared using the same method as described above and tested individually to determine reproducibility. The degrees of similarity of the five samples calculated by the Similarity Degree Calculation Software for Traditional Chinese Medicine Fingerprints were all above 0.98, indicating that the sample preparation method was reproducible.

Sample Analysis

Ten batches of samples were analyzed under the conditions described above.

Establishment of Fingerprints

The number of peaks, peak area and relative retention time of the samples from the 10 batches were analyzed and compared to establish an optimized fingerprint (Figs. 40.1–40.3).

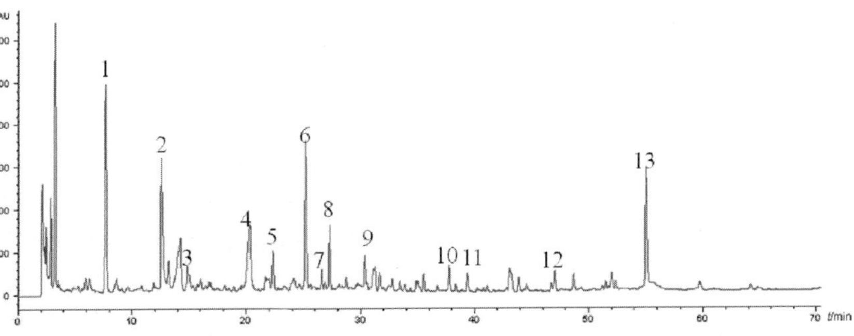

Fig. 40.1. HPLC fingerprint of Yinchaihu (*Radix stellariae*).

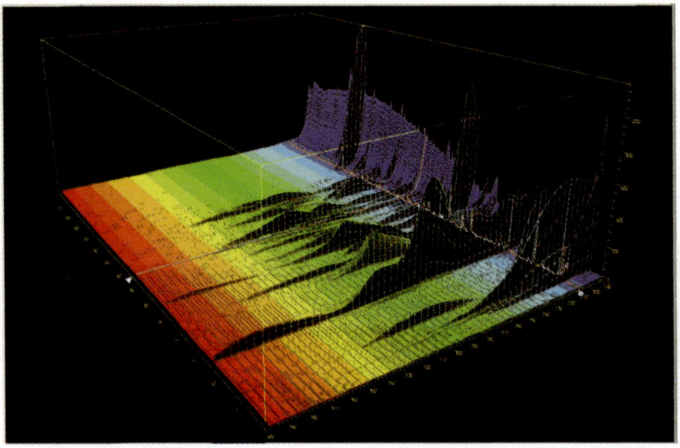

Fig. 40.2. 3-D HPLC fingerprint of Yinchaihu (*Radix stellariae*).

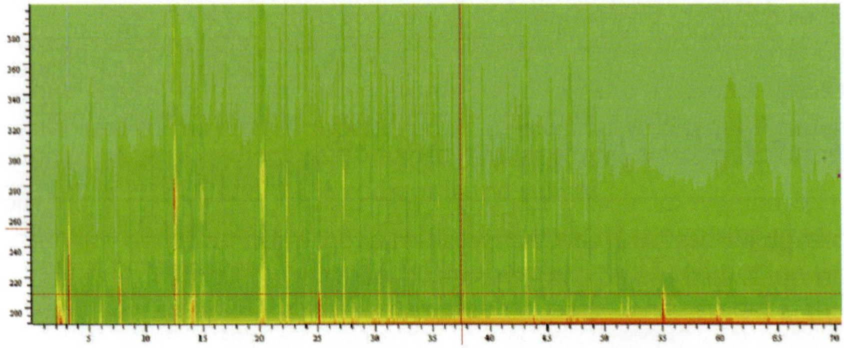

Fig. 40.3. High-resolution HPLC fingerprint of Yinchaihu (*Radix stellariae*).

Results and Discussion

About 40 peaks were separated under the above HPLC conditions. Thirteen common peaks were selected to characterize the fingerprint of *Radix stellariae* by comparing all peaks obtained. The retention time (min) and peak area values of 13 common peaks are listed below.

Peak number	1	2	3	4	5	6	7
Retention time	7.653	12.565	14.905	20.216	22.301	25.14	26.537
Peak area	4690.1	5088.1	1625.9	4610.2	1520.6	4221.2	849.6

Peak number	8	9	10	11	12	13	
Retention time	27.226	30.315	37.724	39.341	48.646	55.015	
Peak area	1524.2	1151.9	944.7	577	886.9	4278.4	

Common Peak Standardization

Common peaks were assigned according to the results of the 10 batches of samples. There were 13 common peaks in the HPLC fingerprint of *Radix stellariae*. Compared to the reference peak (peak 6), the relative retention times of the other 12 common peaks were 0.304, 0.500, 0.593, 0.804, 0.887, 1.056, 1.083, 1.206, 1.501, 1.565, 1.935 and 2.188.

Ratio of Common Peak Area

None of the common peak areas exceeded 10% of the total peak area.

2.41

Xuanfuhua

旋覆花

Flos inulae

Introduction

Flos inulae is the dried capitulum of *Inula japonica* Thunb. or *Inula britannica* L. (family Asteraceae).

It is mainly produced in Henan, Hebei, Jiangsu and Zhejiang Provinces. According to studies on resources, *Flos inulae* is grown in abundance in Anguo and Xianghe of Hebei Province.

Chemical Composition

Flos inulae contains inulicin. *Inula britannica* L. contains gaillardin, quercetin, quercetin glycoside, isoquercetin and quercetagetin glycoside.

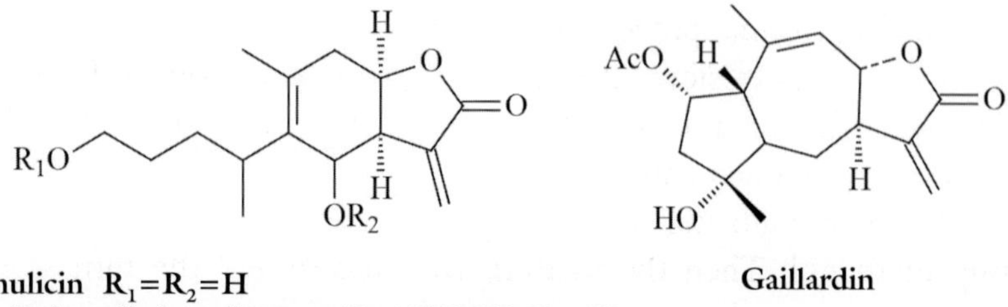

Inulicin $R_1=R_2=H$
1-O-Acetylbritannilactone $R_1=CH_3CO$, $R_2=H$
1,6-O,O-Diacetylbritannilactone $R_1=R_2=CH_3CO$

Gaillardin

Experimental Materials

Samples

Flos inulae was collected from Anguo and Xianghe of Hebei Province, and purchased from the medicinal companies and markets in Pingyao of Shanxi, Kunming of Yunnan, Wulumuqi of Xinjiang, Sanming of Fujian, Yuncheng of Shanxi, Tianjin, Heilongjiang, Shanghai and Hubei from July 2000 to May 2001. There are 12 batches of samples identified as *Flos inulae*.

References

Inulicin and deacetylinulicin were extracted and purified by our lab. The structures of the two references were identified by UV, IR, EI-MS, ^{1}H-NMR, ^{13}C-NMR, ^{1}H-^{1}HNMR, etc. Their purities were both above 99.5% (calculated by the normalization method).

Reagents

Chromatographic-grade acetonitrile was purchased from Fisher Scientific Company, USA. Ultra-pure water was provided by the Medicine and Hygiene Analytical Center in School of Pharmaceutical Sciences at Peking University and was filtered using a 0.45 μm membrane filter.

Preparation of Samples

2 g of *Flos inulae* powder was accurately weighed after passing through a 20 mesh sieve. It was placed in a round-bottom flask and stoppered after 25 ml of methanol was accurately transferred. The mixture was weighed and cooled in cold water for 1 h, then sonicated for 20 min before being weighed again. The weight loss was made up using methanol. Then the mixture was filtered and the filtrate was used as a test sample.

Preparation of Reference

0.2 mg of inulicin and 0.06 mg of deacetylinulicin were accurately weighed and made up to 1 ml with methanol. It was then ready to be used as a reference.

HPLC Conditions

Instrument: Agilent 1100 high-performance liquid chromatography and diode array detector
Mobile phase: H_2O and acetonitrile with gradient elution
Flow rate: 1.0 ml/min
Column: Ultrasphere ODS (4.6 mm × 25 cm, 5 μm) analytical column and ODS (4.6 mm × 4.5 cm, 5 μm) guard column
Column temperature: 40°C
Detecting wavelength: 210 nm
Run time: 90 min
Injection volume: 10 μl

Methods

A proper amount of sample solution was injected into the liquid chromatograph. The chromatogram was obtained under selected HPLC conditions. The experimental method was validated.

Mobile Phase Selection

A comparison of experimental results showed that gradient elution with water (A) and acetonitrile (B) could successfully separate the peaks and produce the maximum number of peaks from the samples. The duration of analysis was 90 min. No characteristic peaks appeared after 90 min in the extended analysis of up to 2 h.

A is water, B is acetonitrile.

	A (%)	B (%)
0–35 min	70–100	0–30
35–50 min	60–70	30–40
50–90 min	0–60	40–100

Method Validation

1. Stability: As the interval for sample injection is 110 min, a stability study of the sample solution was carried out. Results showed that the absolute retention time, relative retention time and ratio of peak area were quite consistent and their respective RSDs were 0.01–0.75%, 0–0.7% and 0.16–2.70%, which complied with testing requirements.
2. Precision: Five consecutive injections of the samples from the same batch were analyzed to determine the degree of precision. The absolute retention time, relative retention time and ratio of peak area were compared for all injections, and results indicated that the injection method was carried out with acceptable precision.
3. Reproducibility: Five samples from the same batch were weighed accurately and prepared using the same method as described above and tested individually to determine reproducibility. The relative retention time and ratio of peak area were calculated. Results showed that the absolute retention time, relative retention time and ratio of peak area were quite consistent and their respective RSDs were 0.01–0.38%, 0–0.34% and 0.67–2.60%, which complied with testing requirements.

Sample Analysis

The HPLC fingerprints of the test samples from 10 batches were analyzed under the conditions described above.

Establishment of Fingerprints

The number of peaks, peak area and relative retention time of the samples from the 10 batches were analyzed and compared to establish an optimized fingerprint (Fig. 41.1).

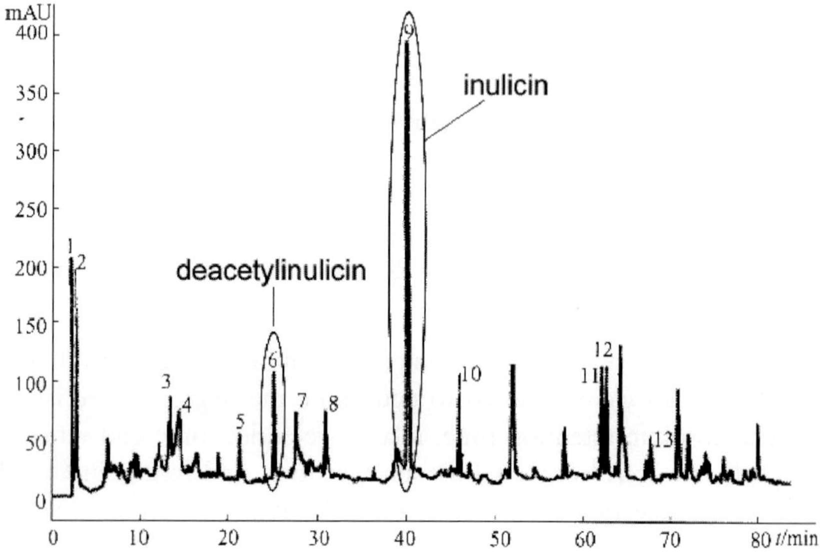

Fig. 41.1. HPLC fingerprint of Xuanfuhua (*Flos inulae*).

Results and Discussion

About 30 peaks were separated under the HPLC conditions described above. Thirteen common peaks were selected to characterize the fingerprint of *Flos inulae*.

Common Peak Standardization

Common peaks were assigned according to the results of the 10 batches of samples. There were 13 common peaks in the HPLC fingerprint of *Flos inulae*. Peak 9 was selected as a reference peak, and the rest of the peaks were compared against the reference peak in terms of retention time. The relative retention times of the remaining 12 common peaks were 0.06, 0.07, 0.38, 0.42, 0.53, 0.63, 0.69, 0.77, 1.15, 1.55, 1.56 and 1.69.

Ratio of Common Peak Area

Aside from the reference peak, no other common peak had an area exceeding 10% of the total peak area.

2.42

Gegen
葛根
Radix puerariae

Introduction

Radix puerariae is the dried root of wild *Pueraria lobata* (Willd.) Ohwi or *Pueraria thomsonii* Benth. Wild *Pueraria lobata* is mainly found in Hunan, Henan, Guangdong, Zhejiang and Sichuan. *Pueraria thomsonii* Benth. is mainly cultivated in Guangdong and Guangxi Provinces. According to studies on resources, wild *Pueraria lobata* grows extremely well and in abundance in Hangzhou and Huzhou of Zhejiang Province.

Chemical Composition

Wild *Pueraria lobata* contains flavonoids. It mainly contains daizein, daidzein, puerarin, daidzein-4′,7-diglucoside, puerarin-7-xyloside, 4′,6′-*O*-diacetyl puerarin, genistein and formononetin, which are derivatives of isoflavone. It also contains kakonein, puerarol, miroestrol and allantoin. Other ingredients include β-sitosterol, daucosterol, 6,7-dimethoxycoumarin, 5-methylhydantoin and amino acids.

The total content of flavonoids in the root of *Pueraria thomsonii* Benth. is lower than that in *Pueraria lobata*. *Pueraria thomsonii* Benth. also contains glycoside of indanol and indirubin indole, β-sitosterol, adenosine, arginine, proline, glutamic acid, γ-aminobutyric acid, leucine and palmitic acid.

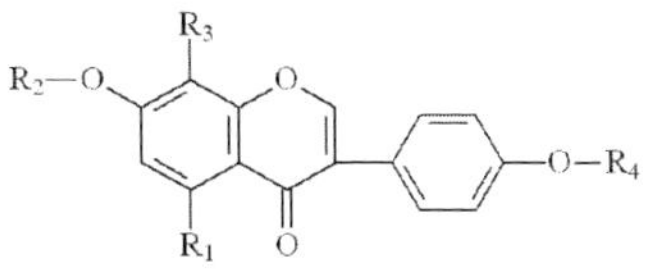

	R_1	R_2	R_3	R_4
Daidzein	H	H	H	H
Daizein	H	glu	H	H
Puerarin	H	H	glu	H
Genistein	OH	H	H	H
Formononetin	H	H	H	CH_3
Daidzein-4′,7-diglucoside	H	glu	H	glu

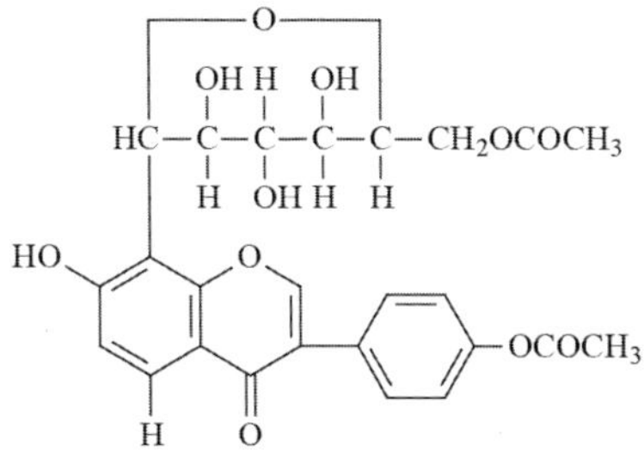

4′,6′-O-Diacetyl puerarin

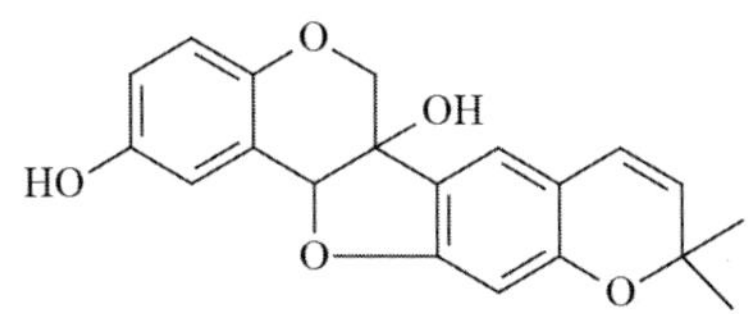

Kakonein

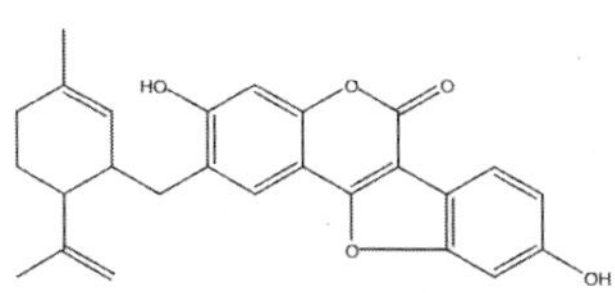

Puerarol

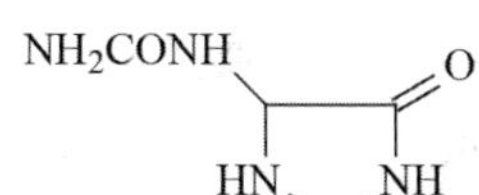

Allantoin

Experimental Materials

Samples

Radix puerariae samples were obtained from Hangzhou and Huzhou of Zhejiang Province in November 2002. There were 12 batches of the sample. They were identified as wild *Pueraria lobata*.

Reference

Puerarin was prepared in our laboratory.

Reagents

Double-distilled deionized water (prepared by our lab). Methanol (analytical-grade methanol was purchased from Shanghai Chemical Reagents Pte. Ltd.; chromatographic-grade methanol was purchased from Tedia Company Inc.). Ethanol (analytical-grade, Shanghai Chemical Reagents Pte. Ltd.).

Preparation of Samples

0.1 g of crude powder of *Radix puerariae* was accurately weighed and placed in a round-bottom flask. 50 ml of 30% ethanol was added and the total weight was taken. The mixture was cooled and weighed after refluxing for 30 min. The weight loss was recovered by addition of 30% ethanol. The mixture was filtered with a 0.45 μm Millipore filter. The filtrate was then ready to be used as a test sample.

Preparation of Reference

1 ml of methanol was added to 0.2 mg of puerarin standard and mixed well to obtain the reference solution.

HPLC Conditions

Instrument: Agilent 1100 high-performance liquid chromatography and diode array detector
Mobile phase: 75% water and 25% methanol
Flow rate: 1.0 ml/min
Column: Agilent ZORBAX 80A Extend-C_{18} analytical column (250 mm × 4.6 mm, 5 μm)
Column temperature: 25°C
Detecting wavelength: 250 nm
Reference wavelength: 360 nm
Run time: 30 min
Injection volume: 20 μl

Methods

A proper amount of the sample was injected into the liquid chromatograph. The chromatogram was obtained under selected HPLC conditions. The experimental method was validated.

Mobile Phase Selection

A comparison of experimental results showed that gradient elution with water and acetonitrile could successfully separate the peaks and produce the maximum number of peaks from the samples. The duration of the analysis was 30 min. The 1.5 h chromatogram of the samples showed no characteristic peaks after 30 min.

Method Validation

1. Stability: Test samples from the same batch were stored at room temperature and injected at regular intervals for testing. The degrees of similarity of all injections calculated by the Similarity Degree Calculation Software for Traditional Chinese Medicine Fingerprints were all above 0.94, which indicated that samples were stable within the 36 testing hours.
2. Precision: Five consecutive injections of the samples from the same batch were conducted to determine the precision. The degrees of similarity calculated by the Similarity Degree Calculation Software for Traditional Chinese Medicine Fingerprints were all above 0.97, which indicated that the injection method was carried out with acceptable precision.
3. Reproducibility: Five samples from the same batch were prepared and tested according to the method described above. The degrees of similarity of the five samples calculated by the Similarity Degree Calculation Software for Traditional Chinese Medicine Fingerprints were all above 0.94, indicating that the sample preparation method was reproducible.

Sample Analysis

The HPLC fingerprints of the test samples from 10 batches were analyzed under the conditions described above.

Establishment of Fingerprints

The number of peaks, peak area and relative retention time of the samples from the 10 batches were analyzed and compared to establish an optimized fingerprint (Figs. 42.1 and 42.2).

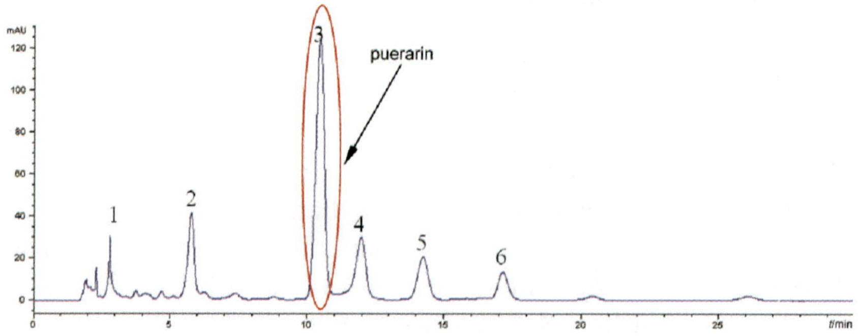

Fig. 42.1. HPLC fingerprint of Gegen (*Radix puerariae*).

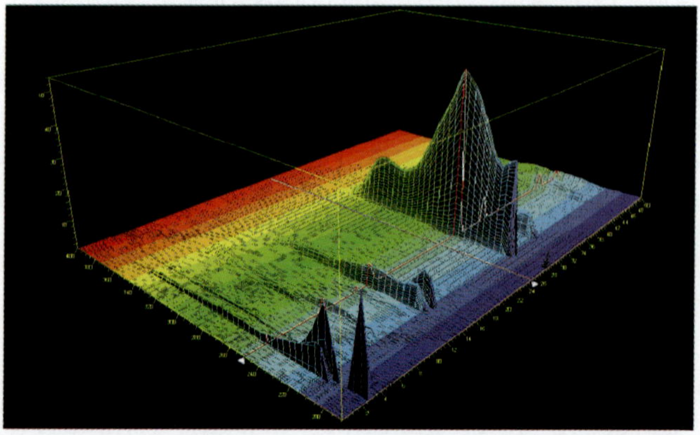

Fig. 42.2. 3-D HPLC fingerprint of Gegen (*Radix puerariae*).

Results and Discussion

About 10 peaks were separated under the above HPLC conditions. Six common peaks were selected to characterize the fingerprint of *Radix puerariae* by comparing all peaks obtained from 10 different batches of samples against the reference peak. The retention time (min) and peak area values of the six common peaks are listed below.

Peak number	1	2	3	4	5	6
Retention time	2.827	5.818	10.488	11.999	14.264	17.164
Peak area	270.9	690.9	2747.6	828.5	553.1	321.8

Common Peak Standardization

Common peaks were assigned according to the results of 10 batches of samples. There were six common peaks in the HPLC fingerprint of *Radix puerariae*. Compared to the reference peak (peak 3), the relative retention times of the other five common peaks were 0.270, 0.555, 1.144, 1.360 and 1.637.

Ratio of Common Peak Area

Aside from the reference peak, the peak areas of peaks 2 and 4 exceeded 10% of the total peak area. The relative peak area ratio of peak 2 was specified to be within 0.1760–0.3269, whilst that for peak 4 was within 0.2111–0.3920.

2.43

Fanxieye
番泻叶
Folium sennae

Introduction

Folium sennae is the small dried leaf of *Cassia angustifolia* Vahl or *Cassia acutifolia* Delile (family Legume).

Cassia angustifolia Vahl is mainly found in areas from the eastern Red Sea to India. Now it is cultivated in abundance in the southern areas of India, in Tinnevelly. It is therefore called Indian or Tinnevelly *Cassia angustifolia* Vahl. It is also produced in Egypt and Sudan. *Cassia acutifolia* Delile is mainly found in the upper-middle areas of the Nile River in Egypt. It is usually exported from the Alexandria Harbor, and is thus called Egyptian or Alexandria *Cassia acutifolia* Delile. It is currently cultured in Guangdong, Hainan Island and Xishuangbanna in Yunnan.

Chemical Composition

The leaf of *Cassia angustifolia* Vahl contains sennosides A, B, C and D. *Adenophora tetraphylla* (Thunb.) Fisch. contains triple terpene saponin, *Radix adenophorae* saponin, β-sitosterol, daucosterol, dandelion terpinoside, saturated fat, aloe-emodin dianthrone glycoside, emodin monoglucoside, aloe-emodin monoglucoside and a small amount of rhein and aloe-emodin. *Cassia acutifolia* Delile contains derivatives of anthraquinone, e.g. sennosides A, B, C, D; aloe-emodin monoglucoside; rhein-8-monoglucoside; rhein-1-monoglucoside and aloe-emodin.

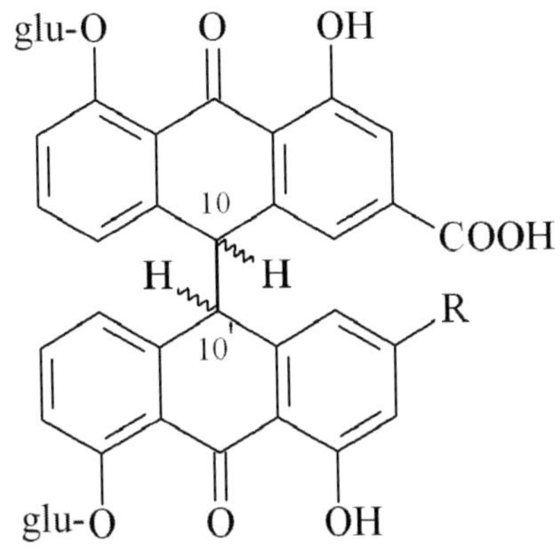

	R	10-10′
Sennoside A	COOH	*trans*
Sennoside B	COOH	*cis*
Sennoside C	CH_2OH	*trans*
Sennoside D	CH_2OH	*cis*

Experimental Materials

Samples

Twelve *Folium sennae* samples were imported from India, and identified as *Cassia angustifolia* Vahl.

Reagents

Acetonitrile (chromatographic-grade, Tedia Company Inc., USA). Methanol (analytical-grade methanol was purchased from Shanghai Chemical Reagents Pte. Ltd.; chromatographic-grade methanol was purchased from Tedia Company Inc.). Double-distilled deionized water (prepared by our lab).

Preparation of Samples

2 g of *Cassia angustifolia* Vahl crude powder was accurately weighed and placed in a round-bottom flask. 100 ml of methanol was added. The mixture was then reflux-extracted for 2 h, and subsequently filtered. The residue was added to 100 ml of methanol, subjected to another 1 h of reflux-extraction and filtered. The filtrates were combined and evaporated to dryness. The residue was dissolved in metha-

nol and topped up to 10 ml. The solution was filtered through a 0.45 μm Millipore filter before injection.

HPLC Conditions

Instrument: Agilent 1100 high-performance liquid chromatography and diode array detector
Mobile phase: H_2O and acetonitrile with gradient elution
Flow rate: 1.0 ml/min
Column: Agilent ZORBAX 80A Extend-C_{18} analytical column (250 mm × 4.6 mm, 5 μm)
Column temperature: 25°C
Detecting wavelength: 240 nm
Reference wavelength: 360 nm
Run time: 70 min
Injection volume: 20 μl

Methods

A proper amount of sample solution was injected into the liquid chromatograph. The chromatogram was obtained under selected HPLC conditions. The experimental method was validated.

Mobile Phase Selection

A comparison of experimental results showed that gradient elution with water (A) and acetonitrile (B) could successfully separate the peaks and produce the maximum number of peaks from the samples. The duration of the analysis was 70 min. No characteristic peaks appeared after 70 min from the extended analysis of up to 2 h.

A is water, B is acetonitrile.

	A (%)	B (%)
0–70 min	6–98	2–94

Method Validation

1. Stability: The sample solution from the same batch was stored at room temperature and injected at regular intervals. The degrees of similarity of all injections calculated by the Similarity Degree Calculation Software for Traditional Chinese Medicine Fingerprints were above 0.95, which indicated that the sample was stable within the 36 testing hours.
2. Precision: Five consecutive injections of the samples from the same batch were analyzed to determine the precision. The degrees of similarity calculated by the Similarity Degree Calculation Software for Traditional Chinese Medicine Fingerprints were all above 0.95, indicating that the injection method was carried out with acceptable precision.
3. Reproducibility: Five samples from the same batch were prepared using the method described above and tested individually to determine reproducibility. The degrees of similarity of the five samples calculated by the Similarity Degree Calculation Software for Traditional Chinese Medicine Fingerprints were all above 0.93, indicating that the sample preparation method was reproducible.

Sample Analysis

Ten batches of samples were analyzed under the same HPLC conditions.

Establishment of Fingerprints

The number of peaks, peak area and relative retention time of the samples from the 10 batches were analyzed and compared to establish an optimized fingerprint (Figs. 43.1 and 43.2).

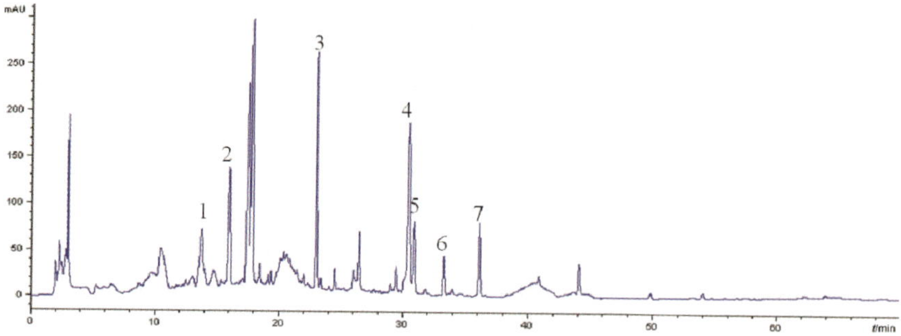

Fig. 43.1. HPLC fingerprint of Fanxieye (*Folium sennae*).

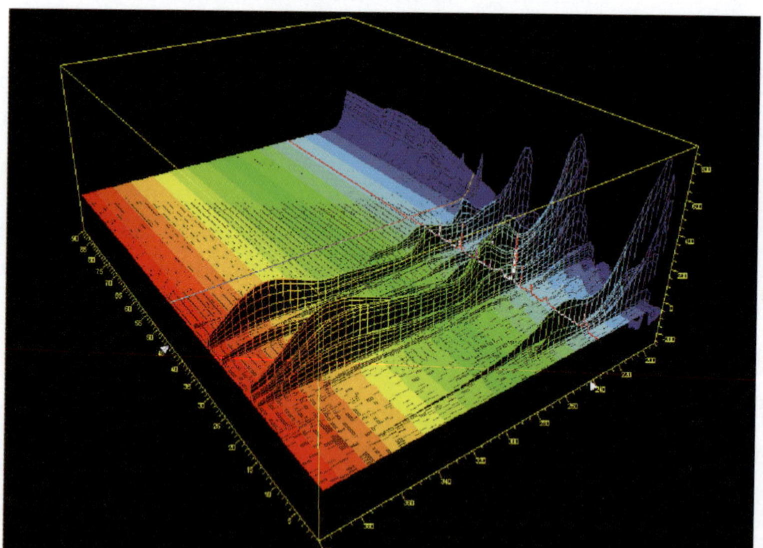

Fig. 43.2. 3-D HPLC fingerprint of Fanxieye (*Folium sennae*).

Results and Discussion

Over 40 peaks were separated under the above HPLC conditions. Seven common peaks were selected to characterize the fingerprint of *Folium sennae* by comparing all peaks obtained. The retention time (min) and peak area values of the seven common peaks are listed below.

Peak number	1	2	3	4	5	6
Retention time	2.827	5.818	10.488	11.999	14.264	17.164
Peak area	270.9	690.9	2747.6	828.5	553.1	321.8

Common Peak Standardization

Common peaks were assigned according to the results of 10 batches of samples. There were seven common peaks in the HPLC fingerprint of *Folium Sennae*. Peak 3 was selected as the reference peak, and the rest of the peaks were compared against the reference peak in terms of retention time. The relative retention times of the other six common peaks were 0.598, 0.696, 1.324, 1.343, 1.447 and 1.571.

Ratio of Common Peak Area

With the exception of the reference peak, no other peak had an area exceeding 10% of the total peak area.

References

1. 国家药典委员会编. (2002) 中华人民共和国药典(2000年版一部). 北京:化学工业出版社.
2. 肖培根主编. (2000) 新编中药志 (第一卷, 第二卷, 第三卷). 北京:化学工业出版社.
3. 王惠清编著. (2004) 中药材产销. 成都:四川出版集团 四川科学技术出版社.
4. 邹忠梅, 徐丽珍, 杨世林. (2003) 芍药总苷高效液相色谱指纹图谱研究.药学学报, **38**(1):46–49.
5. 赵燕燕, 贾凌云, 孙启时. (2002) 白茅根药材的指纹图谱.沈阳药科大学学报, **19**(5):352–354.
6. 赵明波, 邓秀兰, 王亚玲, 等. (2004) 红花RP-HPLC指纹图谱的建立及其质量研究.药学学报, **39**(3):212–216.
7. 粟晓黎, 林瑞超. (2003) 中药侧叶黄酮成分指纹图谱分析方法的研究. 药物分析杂志, **23**(6):417–420.
8. 许勇, 李发美, 郭兴杰. (2003) 补骨脂药材指纹图谱分析方法研究. 中国药学杂志, **38**(7):503–506.
9. 马翠英, 戴宝成, 林瑞超. (2003) 黄芩中4种黄酮的HPLC定量分析及其黄酮类成分指纹图谱研究. 药物分析杂志, **23**(2):83–85.
10. 侯艳鹏, 王建华, 尚明英, 等. (2002) 旋覆花药材HPLC指纹图谱研究. 中国药学杂志, **37**(12):894–899.

Appendix I

Relevant Fingerprint Software

1. Similarity Degree Calculation Software for Traditional Chinese Medicine Fingerprints (provided by Professor Cheng Yiyu of Zhejiang University.
   This software is mainly used for calculation and analysis of fingerprint similarity. It provides five basic functional modules for users:
   1. import data;
   2. pre-process data, resize and compare fingerprint peaks;
   3. retention time correction and auto peak matching;
   4. fingerprint similarity calculation;
   5. export results and print report.

2. Computer-assisted Similarity Evaluation System (provided by Professor Liang Yizeng of Central South University). This software is mainly used for calculation and analysis of fingerprint similarity. It provides four basic functional modules for users:
   1. pre-process data;
   2. pattern analysis of fingerprint;
   3. process fingerprint peaks;
   4. similarity calculation.

The included angle cosine method is used for similarity calculation in the above software. Every fingerprint is regarded as a group of peak area values (or peak height) with their corresponding retention time. If this group of values is regarded as vectors in a multi-dimensional space, the similarity of two fingerprints is now converted into the similarity of two vectors in a multi-dimensional space. The cos θ value is used for quantitative characterization of the similarity between fingerprints. The closer cos θ is to 1, the more similar the two vectors are.

Appendix II

Fingerprints of Injection Formulations and Intermediates

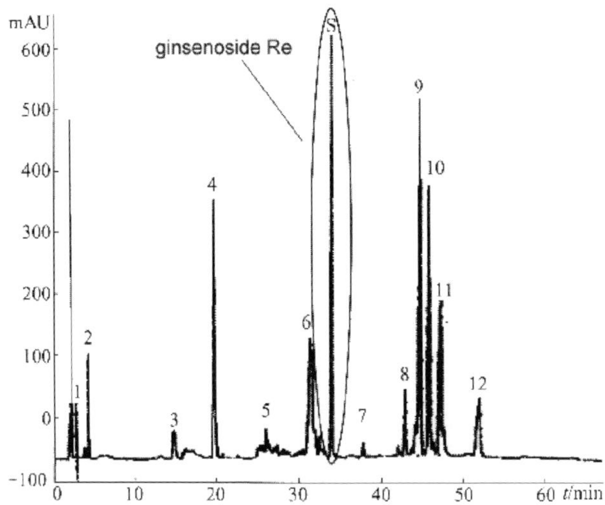

Fig. A1. HPLC fingerprint of intermediates in Renshen.
(Control: ginsenoside Re)

Column: ZORBAX SB-C_{18} (4.6 mm x 250 mm, 5 μm)	Column temperature: 30°C
Mobile phase: Acetonitrile : water (gradient elution)	Detecting wavelength: 203 nm
Flow rate: 1.0 ml/min	Run time: 77 min

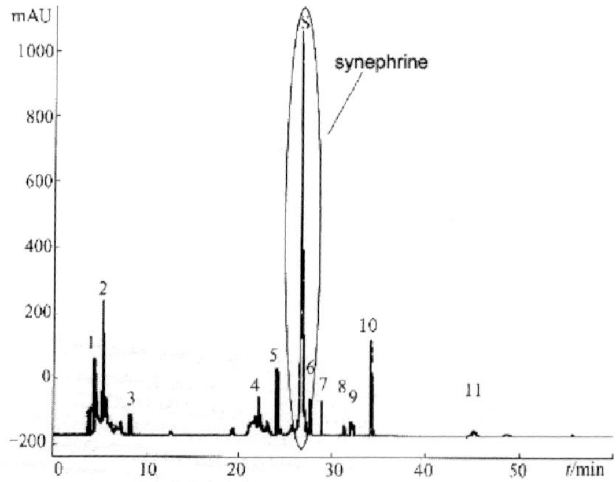

Fig. A2. HPLC fingerprint of intermediates in Chenpi.
(Control: synephrine)

Column: ZORBAX SB-C_{18} (4.6 mm × 250 mm, 5 μm)	Column temperature: 30°C
Mobile phase: Gradient elution with methanol and water (which contains 0.02% H_3PO_4 and 0.1% sodium lauryl sulfate)	Detecting wavelength: 224 nm
Flow rate: 0.6 ml/min	Run time: 60 min

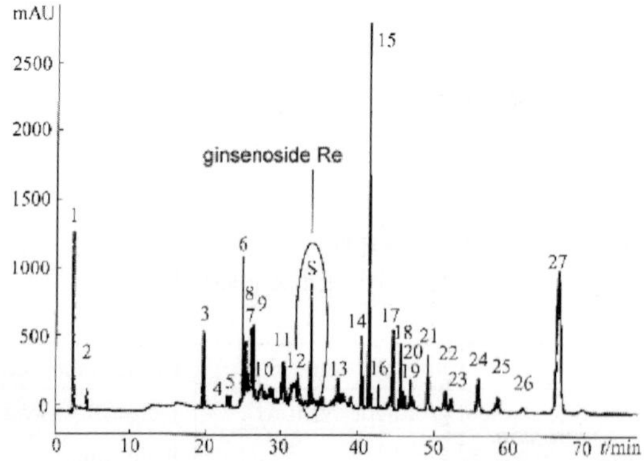

Fig. A3. HPLC fingerprint of intermediates in compound ginseng injection.
(Control: ginsenoside Re)

Column: ZORBAX SB-C_{18} (4.6 mm × 250 mm, 5 μm)	Column temperature: 30°C
Mobile phase: Acetonitrile : water (gradient elution)	Detecting wavelength: 203 nm
Flow rate: 1.0 ml/min	Run time: 77 min

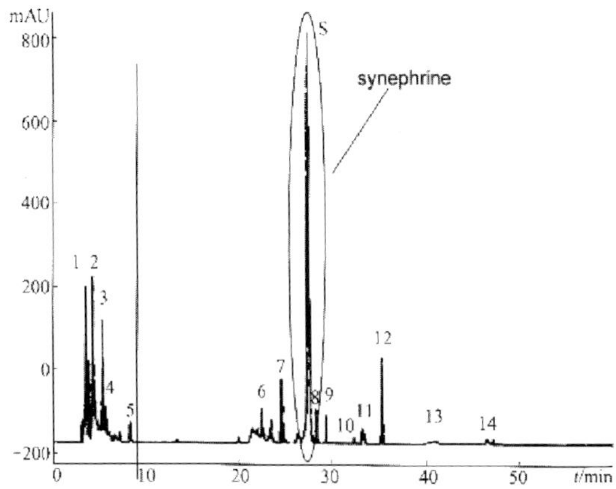

Fig. A4. HPLC fingerprint of intermediates in compound ginseng injection. (Control: synephrine)

Column: ZORBAX SB-C_{18} (4.6 mm × 250 mm, 5 μm)	Column temperature: 30°C
Mobile phase: Gradient elution with methanol and water (which contains 0.02% H_3PO_4 and 0.1% sodium lauryl sulfate)	Detecting wavelength: 224 nm
Flow rate: 0.6 ml/min	Run time: 60 min

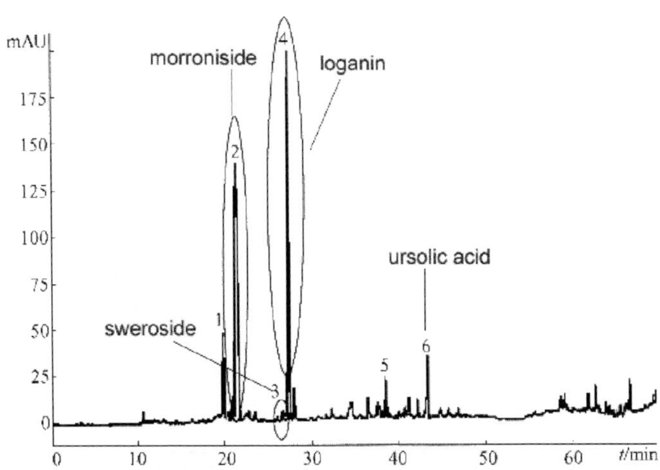

Fig. A5. HPLC fingerprint of intermediates in Shanzhuyu preparation. (Controls: loganin, ursolic acid, morroniside, sweroside)

Column: Phenomenex® Luna C_{18} analytical column (250 mm × 4.6 mm, 5 μm)	Detecting wavelength: 240 nm
Mobile phase: Gradient elution with 0.1% phosphoric acid in water and 0.1% phosphoric acid in acetonitrile	Reference wavelength: 360 nm
Flow rate: 1.0 ml/min	Run time: 70 min
Column temperature: 30°C	

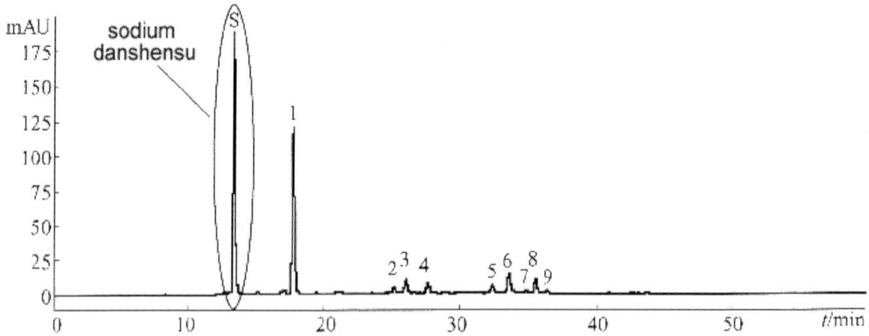

Fig. A6. HPLC fingerprint of intermediates in Danshen. (Control: sodium danshensu)

Column: Hypersil ODS (4.6 mm × 250 mm, 5 μm)	Column temperature: 30°C
Mobile phase: Gradient elution with methanol and 0.5% glacial acetic acid	Detecting wavelength: 279 nm
Flow rate: 0.6 ml/min	Run time: 60 min

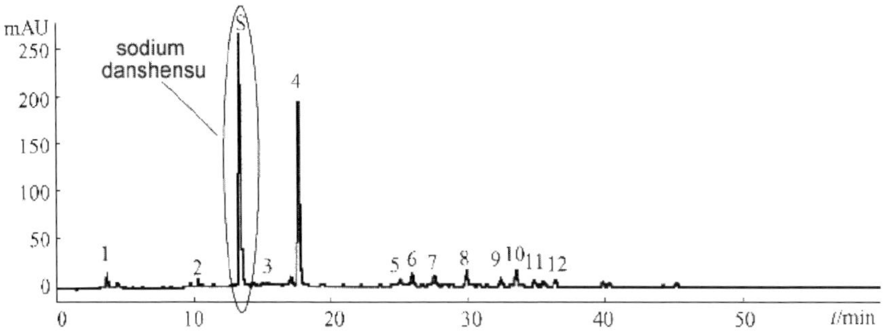

Fig. A7. HPLC fingerprint of Danshen injection formulation.
(Control: sodium danshensu)

Column: Hypersil ODS (4.6 mm × 250 mm, 5 μm)	Column temperature: 30°C
Mobile phase: Gradient elution with methanol and 0.5% glacial acetic acid	Detecting wavelength: 279 nm
Flow rate: 0.6 ml/min	Run time: 60 min

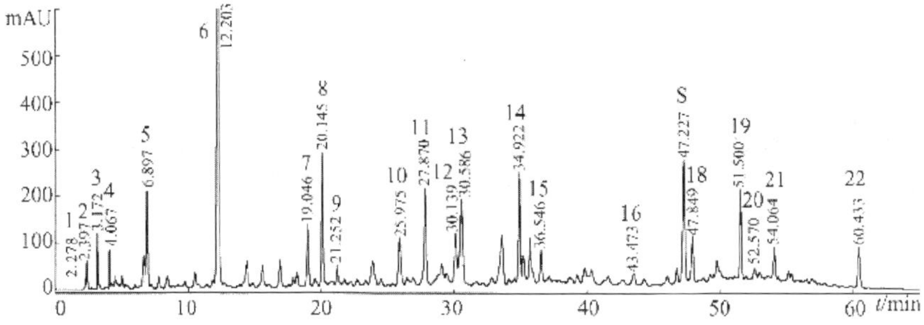

Fig. A8. HPLC fingerprint of oral antiviral decoction.

Column: Hydrosphere C_{18}, S-5 μm (12 nm, 250 mm × 4.6 mm)	Column temperature: 30°C
Mobile phase: Gradient elution with acetonitrile and 0.01% phosphoric acid	Detecting wavelength: 236 nm
Flow rate: 1.0 ml/min	Run time: 65 min

A is acetonitrile, B is 0.01% phosphoric acid.

	A (%)	B (%)
0–22 min	7–18	82–93
22–29 min	18–8	82–82
29–31 min	18–23	77–82
31–40 min	18–23	77–82
40–53 min	23–40	60–77
53–60 min	23–40	60–77
60–65 min	7–40	60–93

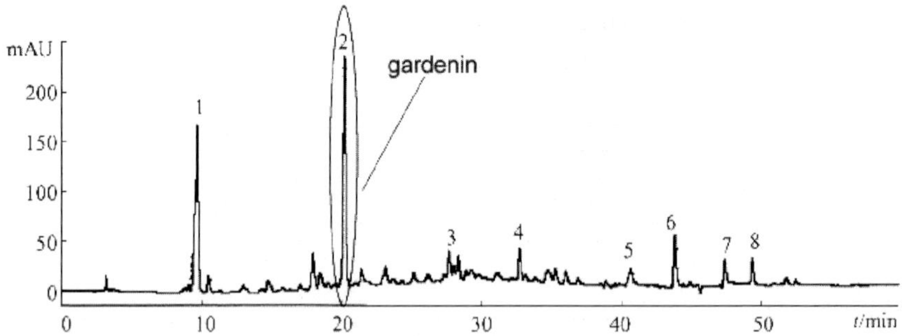

Fig. A9. HPLC fingerprint of Naoluoqing intermediates. (Control: gardenin)

Column: Inertsil® ODS-3 (250 mm × 4.6 mm, 5 μm)	Detecting wavelength: 248 nm
Mobile phase: Gradient elution with 0.1% phosphoric acid in water and 0.1% phosphoric acid in methanol	Reference wavelength: 360 nm
Flow rate: 1.0 ml/min	Run time: 60 min
Column temperature: 25°C	

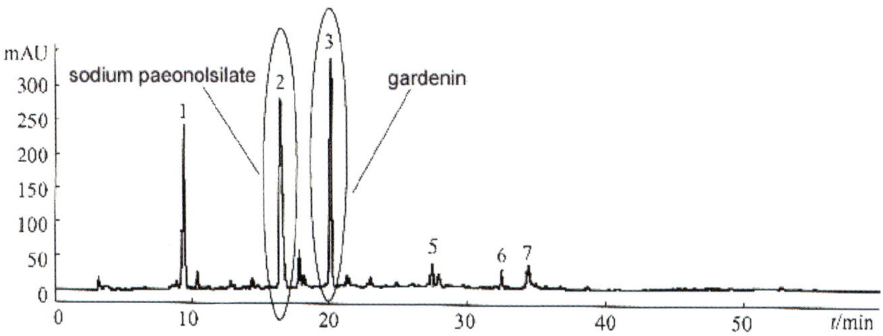

Fig. A10. HPLC fingerprint of Naoluoqing injection.
(Controls: gardenin, sodium paeonolsilate)

Column: Inertsil® ODS-3 (250 mm × 4.6 mm, 5 μm)	Detecting wavelength: 248 nm
Mobile phase: Gradient elution with 0.1% phosphoric acid in water and 0.1% phosphoric acid in methanol	Reference wavelength: 360 nm
Flow rate: 1.0 ml/min	Run time: 60 min
Column temperature: 25°C	

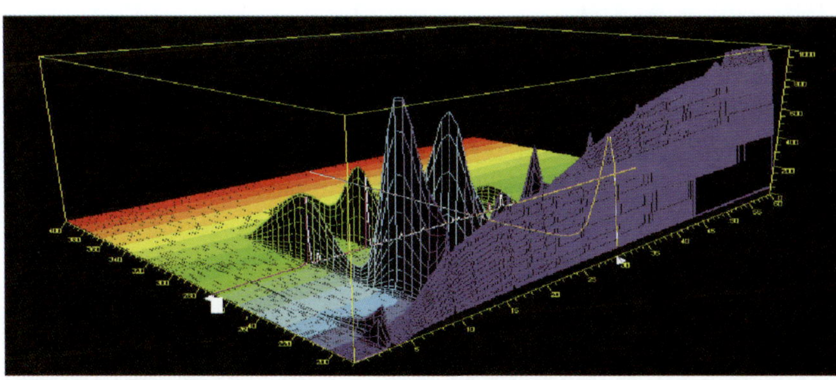

Fig. A11. 3-D HPLC fingerprint of Naoluoqing injection.

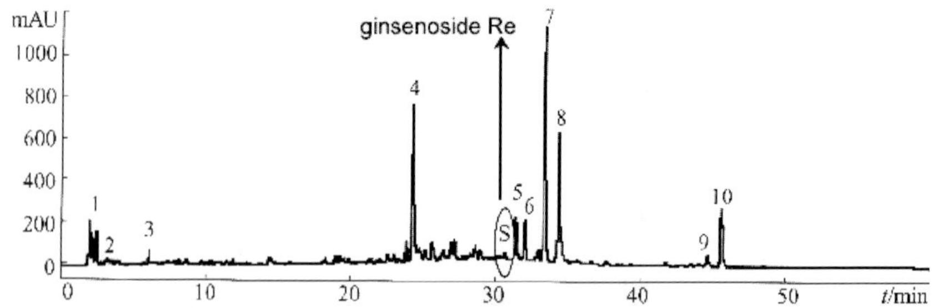

Fig. A12. HPLC fingerprint of intermediates in Huangqi formulation. (Internal standard: ginsenoside Re)

Column: Hypersil ODS (4.6 mm × 250 mm, 5 μm)	Column temperature: 30°C
Mobile phase: Gradient elution with water and acetonitrile	Detecting wavelength: 203 nm
Flow rate: 1.0 ml/min	Run time: 60 min

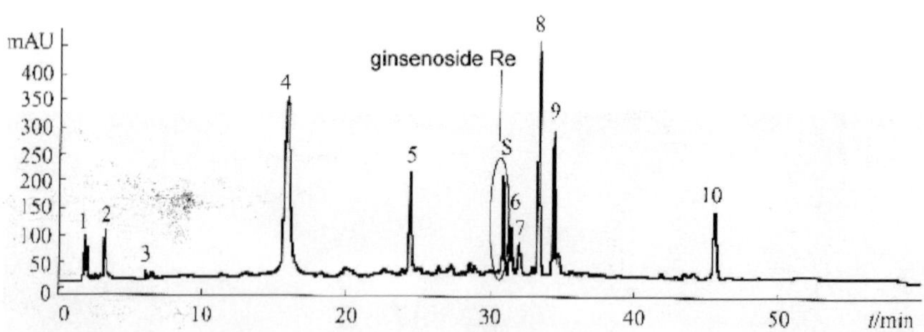

Fig. A13. HPLC fingerprint of Huangqi injection formulation. (Internal standard: ginsenoside Re)

Column: Hypersil ODS (4.6 mm × 250 mm, 5 μm)	Column temperature: 30°C
Mobile phase: Gradient elution with water and acetonitrile	Detecting wavelength: 203 nm
Flow rate: 1.0 ml/min	Run time: 60 min

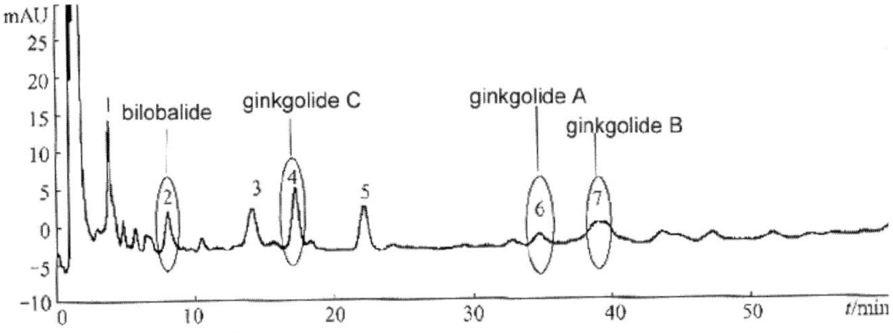

Fig. A14. HPLC fingerprint of Yinxinye extracts.
(Controls: bilobalide, ginkgolides A, B, C)

Column: Hypersil ODS-3 (125 mm × 4.0 mm, 5 μm)	Detecting wavelength: 220 nm
Mobile phase: Gradient elution with double-distilled water and methanol	Reference wavelength: 360 nm
Flow rate: 1.0 ml/min	Run time: 75 min
Column temperature: 30°C	

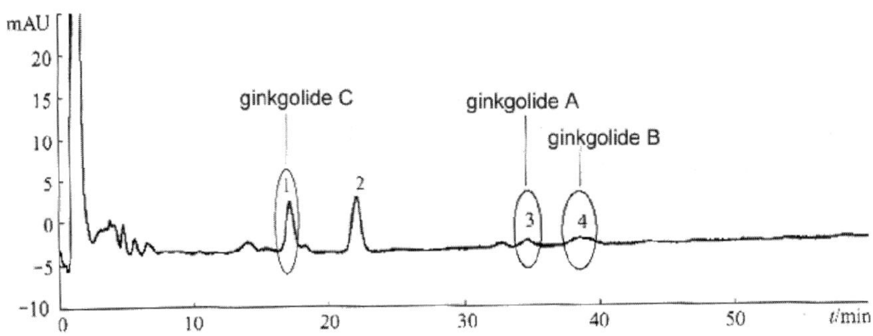

Fig. A15. HPLC fingerprint of Yinxinye injection.
(Controls: ginkgolides A, B, C)

Column: Hypersil ODS-3 (125 mm × 4.0 mm, 5 μm)	Detecting wavelength: 220 nm
Mobile phase: Gradient elution with double-distilled water and methanol	Reference wavelength: 360 nm
Flow rate: 1.0 ml/min	Run time: 75 min
Column temperature: 30°C	

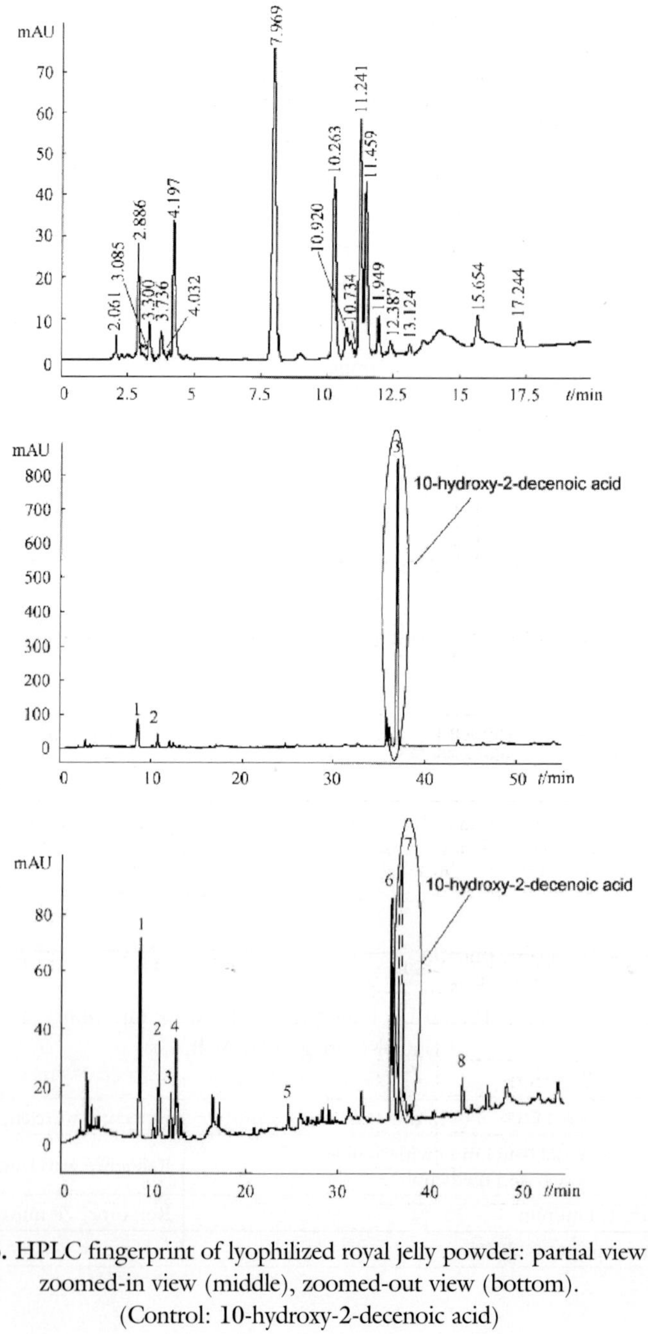

Fig. A16. HPLC fingerprint of lyophilized royal jelly powder: partial view (top), zoomed-in view (middle), zoomed-out view (bottom). (Control: 10-hydroxy-2-decenoic acid)

Column: Phenomenex® Luna C_{18} (4.6 mm × 250 mm, 5 μm)	Detecting wavelength: 230 nm
Mobile phase: Gradient elution with 0.1% phosphoric acid in water and 0.1% phosphoric acid in methanol	Reference wavelength: 360 nm
Flow rate: 1.0 ml/min	Run time: 55 min
Column temperature: 25°C	

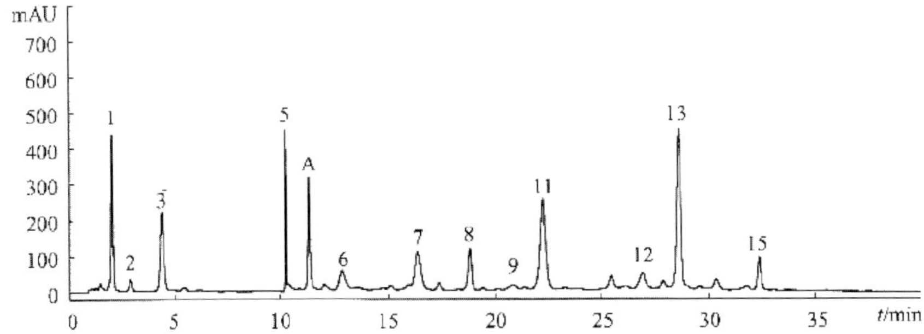

Fig. A17. HPLC fingerprint of Zangqingguo lozenges.

Column: LiChrospher 100 RP-18 (4 mm × 125 mm, 5 μm) (Merck) or other ODS column which has similar properties and capacities	Flow rate: 1.0 ml/min
	Column temperature: 20°C
Mobile phase: Gradient elution with 0.05 mol/L phosphoric acid and 0.05 mol/L potassium dihydrogen phosphate (1:1), methanol and acetic ether	Detecting wavelength: 280 nm
	Run time: 40 min

A is 0.05 mol/L phosphoric acid and 0.05 mol/L potassium dihydrogen phosphate (1:1), B is methanol, C is ethyl acetate.

	A (%)	B (%)	C (%)
0–5 min	94–96	3–6	0–1
5–15 min	93–96	2–3	1–5
15–20 min	89–93	2–6	5–5
20–35 min	55–89	6–40	5–5

Index

L

M

N

P

Q

R

S

T